Life Cycle Costing
for Engineers

Life Cycle Costing
for Engineers

B.S. DHILLON

CRC Press
Taylor & Francis Group
Boca Raton   London   New York

CRC Press is an imprint of the
Taylor & Francis Group, an **informa** business

CRC Press
Taylor & Francis Group
6000 Broken Sound Parkway NW, Suite 300
Boca Raton, FL 33487-2742

First issued in paperback 2017

ISBN 13: 978-1-4398-1688-2 (hbk)
ISBN 13: 978-1-138-07202-2 (pbk)

This book contains information obtained from authentic and highly regarded sources. Reasonable efforts have been made to publish reliable data and information, but the author and publisher cannot assume responsibility for the validity of all materials or the consequences of their use. The authors and publishers have attempted to trace the copyright holders of all material reproduced in this publication and apologize to copyright holders if permission to publish in this form has not been obtained. If any copyright material has not been acknowledged please write and let us know so we may rectify in any future reprint.

Trademark Notice: Product or corporate names may be trademarks or registered trademarks, and are used only for identification and explanation without intent to infringe.

Library of Congress Cataloging-in-Publication Data

Dhillon, B. S. (Balbir S.), 1947-
    Life cycle costing for engineers / author, editor, B.S. Dhillon.
       p. cm.
    "A CRC title."
    Includes bibliographical references and index.
    ISBN 978-1-4398-1688-2 (hard back : alk. paper)
    1. Life cycle costing. 2. Engineering economy. 3. Product life cycle. I. Title.

TA177.7.D3525 2010
658.15'52--dc22
                                         2009030894

Visit the Taylor & Francis Web site at
http://www.taylorandfrancis.com

and the CRC Press Web site at
http://www.crcpress.com

This book is affectionately dedicated to my dear friend,

Dr. G. S. Guram, in thanks for his guidance, honesty, support,

and friendship over the years.

Contents

Preface

Today, in the global economy, the procurement decisions for many engineering products, particularly expensive ones, are not made on initial procurement costs alone, but rather on their life cycle costs. Past experiences indicate that often product ownership cost exceeds the procurement cost. In fact, according to some studies, the product ownership cost (i.e., logistics and operating cost) can vary from 10 to 100 times the original acquisition cost.

Over the past 20 years, a large number of journal and conference proceedings articles on life cycle costing have appeared; however, to my knowledge, only two or three books specifically covering certain areas of civil engineering have been published. More specifically, no general book on life cycle costing was published during this period. In 1989, I published a general book on the topic by reviewing and listing all the journal and conference proceedings articles up to 1989.

The absence of an up-to-date general book on the topic has caused a great deal of difficulty for information seekers because they have had to consult many different and diverse sources. Thus, the main objective of this book is to cover all the latest and most useful aspects of life cycle costing in a single volume and thus eliminate the need to consult many different and diverse sources to obtain desired information. The sources of most of the material presented are listed in the reference section at the end of each chapter. These will be useful to readers who desire to delve more deeply into a specific area or topic.

The book contains a chapter on life cycle costing economics and another on introductory engineering reliability and maintainability concepts considered useful to understanding other chapters of the book. The topics covered in the book are treated in such a manner that the reader does not need previous knowledge to understand the contents. At appropriate places, the book contains examples, along with their solutions; at the end of each chapter, numerous problems test reader comprehension. An extensive list of publications on life cycle costing covering the period from 1988 to 2008 is provided in the bibliography at the end of this book to give readers a view of the intensity of developments on the topic.

The book is composed of 11 chapters. Chapter 1 presents a historical background of life cycle costing, frequently used terms and definition in life cycle costing, useful information on life cycle costing, and the scope of the book. Chapter 2 is devoted to economics concepts considered useful to perform life cycle cost analysis; it also covers topics such as simple interest, compound interest, effective annual interest rate, time-dependent economics formulas, and depreciation methods.

Chapter 3 presents various aspects of life cycle costing fundamentals, including the need and information required for life cycle costing, life cycle costing application areas, approach for incorporating life cycle costing into the planning process for proposals and contracts, areas for evaluating the life cycle costing program, life cycle costing advantages and disadvantages, and life cycle costing data sources. A number of life cycle cost models and cost estimation methods are covered in Chapter 4. The life cycle cost models in the chapter are divided into two areas: general and specific.

Chapter 5 is devoted to reliability, quality, safety, and manufacturing costing. Some of the topics covered in the chapter are reliability cost classifications, models for estimating the cost of reliability-related tasks, quality cost classifications, quality cost indexes, safety cost and its related facts and figures, safety cost estimation models, and manufacturing cost estimation models. Chapter 6 presents various important aspects of maintenance, maintainability, usability, and warranty costing. It covers topics such as reasons for maintenance costing, factors influencing maintenance cost, types of maintenance costs, preventive and corrective maintenance labor cost estimation, maintenance cost data collection, maintainability investment cost elements, manufacturer warranty and reliability improvement warranty costs, usability costing and related facts and figures, and principal costs of ignoring product usability and product usability cost estimation.

Chapters 7 and 8 are devoted to computer system life cycle costing and transportation system life cycle costing, respectively. Some of the topics covered in Chapter 7 are computer system life cycle cost models, computer system maintenance cost, software life cycle cost influencing factors and model, and software cost estimation methods and models. Chapter 8 includes topics such as aircraft life cycle cost, aircraft turbine engine life cycle cost, aircraft cost drivers, cargo ship life cycle cost, ship operating and support costs, urban rail life cycle cost, and city bus life cycle cost estimation models.

Chapter 9 presents various important aspects of civil engineering structures and energy systems life cycle costing. Some of the topics covered in the chapter are building life cycle cost, steel structure life cycle cost, bridge and waste treatment facilities life cycle costs, building energy cost estimation, appliance life cycle costing, and an energy system life cost estimation model. Chapter 10 is devoted to miscellaneous cost estimation models and it presents a total of 12 such models. Some of these models include the plant cost estimation model, program error cost estimation model, satellite procurement cost estimation model, and tank gun system life cycle cost estimation model.

Finally, Chapter 11 presents various introductory aspects of engineering reliability and maintainability. The topics covered in the chapter include bathtub hazard rate curve; common reliability networks; general reliability, mean time to failure, and hazard rate formulas; maintainability measures; and reliability and maintainability tools.

This book will be useful to many individuals, including engineering professionals at large, engineering undergraduate and graduate students, engineering administrators, cost analysts, engineering researchers and instructors, and procurement professionals.

I am deeply indebted to many individuals, including colleagues, students, and friends, for their input and encouragement throughout the project. I thank my children, Jasmine and Mark, for their patience, as well as intermittent disturbances that resulted in many desirable breaks! Last, but not least, I thank my boss, other half, and wife, Rosy, for typing various portions of this book and for her timely help in proofreading and tolerance.

B. S. Dhillon
Ottawa, Ontario

The Author

B. S. Dhillon is a professor of engineering management in the Department of Mechanical Engineering at the University of Ottawa. He has served as a chairman/director of the Mechanical Engineering Department/Engineering Management Program for over 10 years at the same institution. He has published 343 articles (201 journal articles and 142 conference proceedings) on reliability, safety, engineering management, etc. He is or has been on the editorial boards of nine international scientific journals. In addition, Dr. Dhillon has written 35 books on various aspects of reliability, design, safety, quality, and engineering management published by Wiley (1981), Van Nostrand (1982), Butterworth (1983), Marcel Dekker (1984), Pergamon Press (1986), etc. His books are being used in over 100 countries and many of them have been translated into languages such as German, Russian, and Chinese. He served as general chairman of two international conferences on reliability and quality control held in Los Angeles and Paris in 1987.

Professor Dhillon has served as a consultant to various organizations and bodies and has many years of experience in the industrial sector. At the University of Ottawa, he has taught reliability, quality, engineering management, design, and related areas for over 29 years. He has also lectured in over 50 countries, including keynote addresses at various international scientific conferences held in North America, Europe, Asia, and Africa. In March 2004, Dr. Dhillon was a distinguished speaker at the Conference/Workshop on Surgical Errors (sponsored by the White House Health and Safety Committee and the Pentagon) held on Capitol Hill.

Professor Dhillon attended the University of Wales, where he received a BS degree in electrical and electronic engineering and an MS degree in mechanical engineering. He received a PhD degree in industrial engineering from the University of Windsor.

1

Introduction

1.1 Background

Today, in the global economy and due to various other market pressures, the acquisition decisions of many engineering systems, particularly the expensive ones, are not made based on initial procurement costs but rather on their life cycle costs. Past experiences indicate that often engineering system ownership costs exceed acquisition costs. In fact, according to various studies [1], the engineering system ownership cost (i.e., logistic and operating cost) can vary from 10 to 100 times the original acquisition cost.

The life cycle cost of a system may be defined simply as the sum of all costs incurred during its life span (i.e., the total of acquisition and ownership costs). The term *life cycle costing* was used for the first time in 1965 in a report entitled "Life Cycle Costing in Equipment Procurement" [2]. This report was prepared by the Logistics Management Institute, Washington, D.C., for the assistant secretary of defense for installations and logistics, U.S. Department of Defense, Washington, D.C.

As a result of this document, the Department of Defense published a series of three guidelines for life cycle costing procurement, entitled "Life Cycle Costing Procurement Guide (Interim)," "Life Cycle Costing in Equipment Procurement—Casebook," and "Life Cycle Costing Guide for System Acquisitions (Interim)" [3–5]. In 1971, the Department of Defense issued Directive 5000.1, entitled "Acquisition of Major Defense Systems," concerning the requirement for life cycle costing procurement for major systems acquisitions [6].

In 1974, the concept of life cycle costing was formally adopted by the state of Florida and, in 1975, a project entitled "Life Cycle Budgeting and Costing as an Aid in Decision Making" was initiated by the Untied States Department of Health, Education, and Welfare [7]. In 1978, the U.S. Congress passed the National Energy Conservation Policy Act, which made it mandatory for every new federal building to be life cycle cost effective [8].

Since 1974, states such as New Mexico, Alaska, Maryland, North Carolina, and Texas have passed legislation that make life cycle cost analysis mandatory in the planning, design, and construction of all state buildings [8]. In 1981, a journal article presented a comprehensive list of publications on life cycle costing [9]. In 1989, Dhillon presented a list of over 500 publications on various aspects of life cycle costing [8].

Since 1989, many people have contributed to the subject of life cycle costing. An extensive list of publications on life cycle costing covering 1988–2007 is presented in the bibliography at the end of this book.

1.2 Terms and Definitions

Many terms and definitions are used in the area of life cycle costing. Some of the frequently used terms and definitions that are directly or indirectly related to life cycle costing include [8,10–15]:

- *Cost* is the amount of money paid or payable for the acquirement of materials, property, or services.
- *Procurement cost* is the total of investment or acquisition costs (nonrecurring and recurring).
- *Ownership cost* is the total of all costs other than the procurement cost during the life span of an item.
- *Life cycle cost* is the sum of all costs incurred during the life span of an item or system (i.e., the total of procurement and ownership costs).
- *Recurring cost* is the cost that recurs periodically during the life span of a project or item.
- *Nonrecurring cost* is the cost that is not repeated.
- *Reliability* is the probability that an item or system will perform its function satisfactorily for the desired period when used according to specified conditions.
- *Maintainability* is the probability that a failed item or system will be restored to its satisfactory working state within a stated total downtime when maintenance action is started per specified conditions.
- *Downtime* is the total time during which the item or system is not in a condition to perform its specified mission or function.
- *Manufacturing cost* is the sum of fixed and variable costs chargeable to the manufacture of a specified item or system.
- *Maintenance* is all scheduled and unscheduled actions necessary to keep an item or system in a serviceable state or restore it to serviceability. It includes inspection, servicing, modification, repair, etc.
- *Repair cost* is the cost of restoring an item, system, or facility to its original performance or condition.

- *Maintenance cost* is the materials and labor expense required for maintaining items in satisfactory use condition.
- *Mean time to repair* is the average or mean time required to repair an item or system.
- *Failure* is the termination of the ability of an item or system to perform its specified function or mission.
- *Failure rate* is the number of failures of an item or system per unit measure of life (e.g., hours).
- *Compound amount* is the future value of money loaned or invested at compound interest.
- *Redundancy* is the existence of more than one means to perform a specified function.
- *Annuity* is a series of equal payments at equal time intervals.
- *Cost model* is an approach, based on programmatic and technical parameters, for calculating concerned costs.
- *Cost estimating relationship* is an equation that relates cost as the dependent variable to one or more independent variables.
- *Useful life* is the length of time an item or system functions within an acceptable level of failure rate.
- *Mission time* is the time during which the item or system is carrying out its stated mission.

1.3 Useful Information on Life Cycle Costing

There are many sources for obtaining, directly or indirectly, life cycle costing–related information. Some of the most useful sources are listed under the following various different categories.

1.3.1 Journals

- *IEEE Transactions on Reliability*
- *Information and Management*
- *Journal of Quality in Maintenance Engineering*
- *International Power Generation*
- *Microelectronics and Reliability*
- *Better Roads*
- *Journal of Infrastructure Systems*
- *International Journal of Production Research*

- *Railway Gazette International*
- *Concrete Engineering International*
- *IEEE Aerospace and Electronic Systems Magazine*
- *Journal of Transportation Engineering*
- *International Journal of Quality and Reliability Management*
- *Defense Management Journal*
- *Transportation Research Record*
- *International Journal of Production Economics*
- *Chemical Engineering*
- *Quality Engineering*
- *IEEE Transactions on Power Delivery*
- *Reliability Engineering and System Safety*
- *Rail International*
- *European Transactions on Electric Power*

1.3.2 Conference Proceedings

- *Proceedings of the Annual Reliability and Maintainability Symposium*
- *Proceedings of the Annual ISSAT International Conference on Reliability and Quality in Design*
- *Proceedings of the Annual Reliability Engineering Conference for the Electric Power Industry*
- *Proceedings of the Annual American Society for Quality Control (ASQC) Conference*
- *Proceedings of the IEEE Annual Conference on Industrial Electronics*
- *Proceedings of the Annual Offshore Technology Conference*
- *Proceedings of the Annual Canadian Society for Civil Engineering Conference*
- *Proceedings of the Annual Petroleum and Chemical Industry Conference*
- *Proceedings of the IEEE Annual Pulp and Paper Industry Technical Conference*
- *Proceedings of the Annual Conference of the Urban and Regional Information Systems Association*

1.3.3 Technical Reports and Manuals

- MIL-HDBK-259 (Navy), "Life Cycle Cost in Navy Acquisitions," Department of Defense, Washington, D.C., April 1983
- MIL-HDBK-276-1 (MC), "Life Cycle Cost Model for Defense Material Systems Data Collection Workbook," Department of Defense, Washington, D.C., February 1984

- NIST Handbook 135, "Life Cycle Costing Manual: For the Federal Energy Management Program," U.S. Department of Energy, Washington, D.C., 1995
- NISTIR 6806, "Project-Oriented Life Cycle Costing Workshop for Energy Conservation in Buildings," U.S. Department of Energy, Washington, D.C., September 2001
- NISTIR-85-3273-21 (Rev. 4/06), "Energy Price Indices and Discount Factors for Life Cycle Cost Analysis," U.S. Department of Energy, Washington, D.C., April 2006
- "Life Cycle Cost Analysis: A Guide for Architects," American Institute of Architects, Washington, D.C., 1977
- D. E. Peterson, "Life Cycle Cost Analysis of Pavements," Transportation Research Board, National Research Council, Washington, D.C., 1985
- H. Hawk, "Bridge Life Cycle Cost Analysis," Transportation Research Board, National Research Council, Washington, D.C., 2003
- D. M. Frangopol and H. Furuta, editors, "Life Cycle Cost Analysis and Design of Civil Infrastructure Systems," Structural Engineering Institute of the American Society of Civil Engineers, Reston, VA, 2001

1.3.4 Books

- B. S. Blanchard, *Design and Manage to Life Cycle Cost*, M/A Press, Portland, OR, 1978
- A. Boussabaine and R. Kirkham, *Whole Life Cycle Costing*, Blackwell Publishing, Oxford, UK, 2004
- J. W. Bull, editor, *Life Cycle Costing for Construction*, Blackie Academic and Professional, Inc., London, 1993
- W. J. Fabrycky and B. S. Blanchard, *Life Cycle Cost and Economic Analysis*, Prentice Hall, Inc., Englewood Cliffs, NJ, 1991
- B. S. Dhillon, *Life Cycle Costing: Techniques, Models, and Applications*, Gordon and Breach Science Publishers, New York, 1989
- D. Hunkeler, K. Lichtenvort, and G. Rebitzer, editors, *Environmental Life Cycle Costing*, CRC Press, Boca Raton, FL, 2008
- M. R. Seldon, *Life Cycle Costing: A Better Method of Government Procurement*, Westview Press, Boulder, CO, 1979
- A. J. Dell'isola and S. J. Kirk, *Life Cycle Costing for Design Professionals*, McGraw–Hill Book Company, New York, 1981
- M. E. Earles, *Factors, Formulas, and Structures for Life Cycle Costing*, Eddins–Earles, Concord, MA, 1981
- R. J. Brown and R. R. Yanuck, *Life Cycle Costing: A Practical Guide for Energy Managers*, Fairmont Press, Inc., Atlanta, GA, 1980

1.3.5 Data Information Sources

- Government Industry Data Exchange Program (GIDEP)
  GIDEP Operations Center
  Naval Weapons Station
  U.S. Department of Navy
  Seal Beach
  Corona, CA 91720

- National Technical Information Center (NTIS)
  5285 Port Royal Road
  Springfield, VA 22151

- Defense Technical Information Center
  DTIC-FDAC
  8725 John J. Kingman Road, Suite 0944
  Fort Belvoir, VA 22060-6218

- Reliability Analysis Center
  Rome Air Development Center
  Griffiss Air Force Base
  Rome, NY 13441-5700

- American National Standards Institute (ANSI)
  11 W. 42nd Street
  New York, NY 10036

- Technical Services Department
  American Society for Quality
  611 W. Wisconsin Avenue
  P.O. Box 3005
  Milwaukee, WI 53201-3005

1.3.6 Organizations

- American Society of Civil Engineers (ASCE)
  1801 Alexander Bell Drive
  Reston, VA 20190-4400.

- Society of Manufacturing Engineers
  One SME Drive
  Dearborn, MI 48121.

- American Society of Mechanical Engineers (ASME)
  Three Park Avenue
  New York, NY 10016-5990.

- American Society of Heating, Refrigeration and Air Conditioning Engineers (ASHRAE)
  1791 Tullie Circle, NE
  Atlanta, GA 30329.

- American Public Power Association
  1875 Connecticut Avenue, NW, Suite 1200
  Washington, D.C. 20009-5715.

- SOLE—The International Society of Logistics
  8100 Professional Place, Suite 111
  Hyattsville, MD 20785-2229.

- Reliability Society, IEEE
  P.O. Box 1331
  Piscataway, NJ

1.4 Scope of the Book

Nowadays, life cycle costing is receiving increasing attention in various sectors of the economy, including government procurements and industry. Over the past two decades, a large number of journal and conference proceedings articles have appeared; however, to the best of the author's knowledge, only two or three books specifically covering certain areas of civil engineering have been published. More specifically, no general book on life cycle costing has been produced during this period.

Professionals and others involved in life cycle costing need up-to-date information on the subject and generally face a great deal of difficulty because they have to consult many different and diverse sources. This book is an attempt to satisfy this specific need. The book is written after reviewing the currently available literature on life cycle costing. Therefore, all the effort was directed to covering important past and present issues in the field.

Previous knowledge is not generally required to understand the material covered in this book because two chapters on fundamental economics and reliability and maintainability basics are provided to give sufficient background to potential readers. The book will find use in many diverse disciplines and will be useful to engineering professionals at large, engineering undergraduate and graduate students, procurement professionals, engineering instructors and researchers, and engineering administrators.

Problems

1. Write an essay on the historical developments in life cycle costing.
2. List at least five sources for obtaining life cycle costing–related information.
3. List at least five books considered important for obtaining life cycle costing-related information.
4. Define the following three terms:
   - life cycle cost
   - ownership cost
   - nonrecurring cost
5. List three of the most important organizations for obtaining life cycle costing-related information.
6. List five important journals from which to obtain life cycle costing-related information.
7. Define the following terms:
   - repair cost
   - maintenance cost
   - procurement cost
8. What is the difference between the terms *maintainability* and *maintenance*?
9. Compare the meanings of the following terms:
   - recurring cost
   - nonrecurring cost
10. What is the difference between the terms *equipment reliability* and *equipment maintainability*?

References

1. Ryan, W. J. 1968. Procurement views of life cycle costing. *Proceedings of the Annual Symposium on Reliability* 164–168.
2. Logistics Management Institute (LMI). 1965. Life cycle costing in equipment procurement. Report no. LMI task 4C-5, Washington, D.C.
3. U.S. Department of Defense. 1970. Life cycle costing procurement guide (interim). Department of Defense guide no. LCC-1, Washington, D.C.
4. U.S. Department of Defense. 1970. Life cycle costing in equipment procurement—Casebook. Department of Defense guide no. LCC–2, Washington, D.C.
5. U.S. Department of Defense. 1973. Life cycle costing guide for system acquisitions (interim). Department of Defense guide no. LCC–3, Washington, D.C.
6. U.S. Department of Defense. 1971. Acquisition of major defense systems. Department of Defense directive no. 5000.1, Washington, D.C.
7. Earles, M. E. 1978. *Factors, formulas, and structures for life cycle costing.* Concord, MA: Eddins–Earles.
8. Dhillon, B. S. 1989. *Life cycle costing: Techniques, models, and applications.* New York: Gordon and Breach Science Publishers.
9. Dhillon, B. S. 1981. Life cycle cost: A survey. *Microelectronics and Reliability* 21:495–511.
10. Humphreys, K. K., and Wellman, P. 1987. *Basic cost engineering.* New York: Marcel Dekker, Inc.
11. Stewart, R. D., and Wyskida, R. M. 1987. *Cost estimator's reference manual.* New York: John Wiley & Sons.
12. Humphreys, K. K. 1984. *Project and cost engineers' handbook.* New York: Marcel Dekker, Inc.
13. Society of Automotive Engineers, Inc. 1987. Aircraft engine life cycle cost. Document no. SP-721, Warrendale, PA.
14. Seldon, M. R. 1979. *Life cycle costing: A better method of government procurement.* Boulder, CO: Westview Press.
15. Brown, R. J., and Yanuck, R. R. 1980. *Life cycle costing.* Atlanta, GA: The Fairmount Press, Inc.

2

Life Cycle Costing Economics

2.1 Introduction

The discipline of economics plays a key role in life cycle costing because, to calculate the life cycle cost of items, various types of economics-related information are required. Life cycle costing requires that all potential costs be calculated by taking into consideration the time value of money. In modern society, interest and inflation rates are utilized to take into consideration the time value of money.

In fact, the concept of interest is not new; its history may be traced back to 2000 BC in Babylon, where interest on borrowed commodities (e.g., grain) was paid in the form of grain or through other possible means [1]. Thus, in a similar manner in modern times, the future value of present dollars will be greater because of earned interest or smaller because of inflation. Similarly, the present value of an amount of money to be received in the future would generally be less.

In life cycle costing, future costs, such as operation and maintenance costs associated with an item, have to be discounted to their present values before adding them to the item's acquisition or procurement cost. Over the years, many formulas have been developed in the area of economics for converting money from one point of time to another. Such formulas are considered indispensable in life cycle costing.

This chapter presents various aspects of economics considered useful in performing life cycle costing studies.

2.2 Simple Interest

This is the simplest form of interest and it means that the interest is paid only on the original amount of money borrowed, rather than on the accrued interest. Thus, the total interest paid on the borrowed amount of money is expressed by

$$I = (P)(n)(i) \tag{2.1}$$

where
   I is total interest.
   P is principal amount (i.e., borrowed).
   n is total number of interest periods (e.g., years).
   i is interest rate per specified period.

The total amount of money, A, at the end of, say, n years is expressed by

$$A = P + I \tag{2.2}$$

By substituting Equation (2.1) into Equation (2.2), we get

$$A = P + (P)(n)(i)$$
$$= P(1 + ni) \tag{2.3}$$

Example 2.1

A company borrowed $300,000 for a period of 3 years at an annual simple interest rate of 5% to procure engineering equipment. Calculate the total amount of money the company has to pay to the lender at the end of 3 years.

By substituting the given data values into Equation (2.3), we get

$$A = (300,000)(1 + (0.05)(3))$$
$$= \$345,000$$

Thus, the total amount of money the company has to pay to the lender at the end of 3 years is $345,000.

2.3 Compound Interest

In this case, the interest earned on principal amount, P, during each interest period is added (at the end of each period) to the principal amount and thereafter begins earning interest itself for the remaining term of the loan or investment. Thus, at the end of the first interest period (e.g., a year) the total amount is expressed by

$$A_1 = P + (P)(i)$$
$$= P(1 + i) \tag{2.4}$$

where A_1 is the total amount at the end of the first interest period.

At the end of the second interest period (e.g., a year), the total amount is expressed by

$$A_2 = A_1 (1+i) \tag{2.5}$$

By substituting Equation (2.4) into Equation (2.5), we obtain

$$A_2 = P(1+i)(1+i)$$
$$= P(1+i)^2 \tag{2.6}$$

where A_2 is the total amount at the end of the second interest period.

Similarly, at the end of the third interest period (e.g., a year), the total amount is expressed by

$$A_3 = A_2 (1+i) \tag{2.7}$$

By substituting Equation (2.6) into Equation (2.7), we get

$$A_3 = P(1+i)^2 (1+i)$$
$$= P(1+i)^3 \tag{2.8}$$

where A_3 is the total amount at the end of the third interest period.

Thus, at the end of the nth interest period (e.g., a year), the total amount is generalized to the following form:

$$A_n = A_{n-1} (1+i)$$
$$= P(1+i)^n \tag{2.9}$$

where

n is number of interest periods (e.g., years).

A_n is total or compound amount at the end of the nth interest period (e.g., a year).

A_{n-1} is principal amount at the beginning of the nth interest period (e.g., a year).

The total compound interest earned after the nth interest period (e.g., a year) is given by

$$I_c = A_n - P \tag{2.10}$$

Example 2.2

Assume that a person deposited $80,000 in a bank for 7 years at annual interest rate of 7%, compounded annually. Calculate the total amount of money at the end of the specified period and the compound interest earned at the end of the same period.

By substituting the given data values into Equations (2.9) and (2.10), we get

$$A_7 = (80,000)(1+0.07)^7$$

$$= \$128,462.52$$

and

$$I_c = (80,000)(1+0.07)^7 - (80,000)$$

$$= \$48,462.52$$

Thus, the total amount of money and the compound interest earned at the end of 7 years are $128,462.52 and $48,462.52, respectively.

2.4 Effective Annual Interest Rate

This interest rate may be described simply as the true annual interest rate because it considers the effect of all compounding during the year. The effective annual interest rate can be calculated by using the following equation [2]:

$$(1+i_e) = \left(1+\frac{i}{m}\right)^m \tag{2.11}$$

where
i_e is effective annual interest rate.
i is annual nominal interest rate.
m is total number of interest periods in a year.

Note that Equation (2.11) is developed by reasoning that the effective interest rate compounded once a year generates the same interest as a nominal interest rate compounded m times in a year. By rearranging Equation (2.11), we get

$$i_e = \left(1+\frac{i}{m}\right)^m - 1 \tag{2.12}$$

Example 2.3

A person deposited $100,000 in a bank at a nominal interest rate of 8% compounded monthly, for 12 months. Estimate the value of the effective annual interest rate.

By substituting the specified data values into Equation (2.12), we get

$$i_e = \left(1 + \frac{0.08}{12}\right)^{12} - 1$$

$$= 1.08299 - 1$$

$$= 0.08299$$

$$= 8.299\%$$

Thus, the value of the effective annual interest rate is 8.299%.

2.5 Time-Dependent Formulas for Application in Life Cycle Cost Analysis

In the published literature, many time-dependent formulas have been developed that can be used to perform life cycle cost analysis. Some of these formulas are presented next.

2.5.1 Single Payment Future Worth Formula

This formula for compound amount was developed earlier in the chapter (in the section on compound interest). Thus, from Equation (2.9), the future worth (compound amount) is

$$W_f = A_n = P(1+i)^n \tag{2.13}$$

where
W_f is future worth or amount (i.e., principal amount plus interest earned).
n is number of interest periods (e.g., years).
P is principal amount.
i is compound interest rate per specified period.

2.5.2 Single Payment Present Value Formula

From Equation (2.13), the present value of a future amount of money is given by

$$V_p = P = \frac{W_f}{(1+i)^n} \tag{2.14}$$

where V_p is the present value.

Example 2.4

Assume that the total operation and maintenance cost of a piece of engineering equipment after its 5-year usage will be $150,000. Calculate the present value of $150,000 if the annual compound interest rate is 6%.

By substituting the given data values into Equation (2.14), we get

$$V_p = \frac{(150,000)}{(1+0.06)^5}$$

$$= \$112,088.7$$

Thus, the present value of the engineering equipment total operation and maintenance cost is $112,088.70.

2.5.3 Uniform Periodic Payment Future Amount Formula

This formula is concerned with determining the future amount at the end of n interest periods (years) of equal payments made at the end of each interest period. All payments are invested at an annual compound interest rate i. The formula is developed next.

At the end of the first year, after the first payment, the future amount is

$$FA_1 = PA \tag{2.15}$$

where

FA_1 is future amount at the end of the first year.
PA is payment made at the end of a year.

At the end of the second year, after the second payment and the interest earned on FA_1, using Equation (2.4) the future amount is

$$FA_2 = PA + FA_1(1+i) \tag{2.16}$$

where

FA_2 is future amount at the end of the second year.
i is annual compound interest rate.

Substituting Equation (2.15) into Equation (2.16) yields

$$FA_2 = PA + PA(1+i) \tag{2.17}$$

At the end of the third year, after the third payment and the interest earned on FA_2, the future amount is

$$FA_3 = PA + FA_2(1+i) \tag{2.18}$$

where FA_3 is the future amount at the end of the third year.

By substituting Equation (2.17) into Equation (2.18), we get

$$FA_3 = PA + PA(1+i) + PA(1+i)^2 \qquad (2.19)$$

At the end of the fourth year, after the fourth payment and the interest earned on FA_3, the future amount is

$$FA_4 = PA + FA_3(1+i) \qquad (2.20)$$

where FA_4 is the future amount at the end of the fourth year.
Using Equation (2.19) in Equation (2.20) yields

$$FA_4 = PA + PA(1+i) + PA(1+i)^2 + PA(1+i)^3 \qquad (2.21)$$

At the end of the nth year, after the nth payment and the interest earned on FA_{n-1}, the future amount is

$$FA_n = PA + PA(1+i) + \cdots + PA(1+i)^{n-2} + PA(1+i)^{n-1} \qquad (2.22)$$

where FA_n is the future amount at the end of the nth year.
Equation (2.22) is a geometric series that can be summed as follows: Multiply both sides of Equation (2.22) by $(1+i)$ to obtain

$$(1+i)FA_n = PA(1+i) + PA(1+i)^2 + \cdots + PA(1+i)^{n-2} + PA(1+i)^n \qquad (2.23)$$

By subtracting Equation (2.22) from Equation (2.23), we get

$$(1+i)FA_n - FA_n = PA(1+i)^n - PA \qquad (2.24)$$

After rearranging Equation (2.24), we obtain

$$FA_n = \frac{PA[(1+i)^n - 1]}{i} \qquad (2.25)$$

Example 2.5
Assume that a person deposits $30,000 at the end of each year for the next 8 years. Calculate the total future amount of the money deposited after the 8-year period, if the annual compound interest rate is 5%.
By substituting the given data values into Equation (2.25), we get

$$FA = (30,000)\left[\frac{(1+0.05)^8 - 1}{0.05}\right]$$

$$= \$286,473.26$$

Thus, the total future amount of the money deposited after the 8-year period is $286,473.26.

2.5.4 Uniform Periodic Payment Present Value Formula

This formula is concerned with determining the present value or worth at the end of n interest periods (years) of equal payments made at the end of each interest period. All payments are invested at an annual compound interest rate i.

The formula is developed as follows: At the end of the first year, after the first payment, the present value of that payment from Equation (2.14) is

$$V_{p1} = \frac{PA}{(1+i)} \tag{2.26}$$

where
V_{p1} is present value of the payment, PA, made at the end of the first year.
i is annual compound interest rate.

At the end of the second year, after the second payment, the present value of that payment from Equation (2.14) is

$$V_{p2} = \frac{PA}{(1+i)^2} \tag{2.27}$$

where V_{p2} is present value of the payment, PA, made at the end of the second year.

Similarly, at the end of the nth year, after the nth payment, the present value of that payment from Equation (2.14) is

$$V_{pn} = \frac{PA}{(1+i)^n} \tag{2.28}$$

where
V_{pn} is present value of the payment, PA, made at the end of the nth year.
n is number of interest periods or years.

Using Equations (2.26)–(2.28), we get the following equation for the present value of all payments:

$$PV = V_{p1} + V_{p2} + \cdots + V_{pn}$$
$$= \frac{PA}{(1+i)} + \frac{PA}{(1+i)^2} + \cdots + \frac{PA}{(1+i)^n} \tag{2.29}$$

Equation (2.29) is a geometric series that can be summed as follows: Multiply both sides of Equation (2.29) by $\frac{1}{(1+i)}$ to obtain

$$\frac{PA}{(1+i)} = \frac{PA}{(1+i)^2} + \frac{PA}{(1+i)^3} + \cdots + \frac{PA}{(1+i)^{n+1}} \tag{2.30}$$

By subtracting Equation (2.29) from Equation (2.30), we obtain

$$\frac{PV}{(1+i)} - PV = \frac{PA}{(1+i)^{n+1}} - \frac{PA}{(1+i)} \tag{2.31}$$

After rearranging Equation (2.31), we get

$$PV = PA \left[\frac{1-(1+i)^{-n}}{i} \right] \tag{2.32}$$

Example 2.6

Assume that a person deposits $50,000 at the end of each year for the next 5 years. Calculate the present value of all payments, if the annual compound interest rate is 4%.

By substituting the given data values into Equation (2.32), we get

$$PV = (50,000) \left[\frac{1-(1+0.04)^{-5}}{0.04} \right]$$

$$= \$222,591.1$$

Thus, the present value of all payments is $222,591.10.

2.5.5 Formulas to Calculate Value of Annuity Payments When Annuity's Present and Future Values Are Given

An annuity is a series of equal payments at equal time intervals. Thus, from Equation (2.25) the value of annuity payments when the future value of the annuity is known is given by

$$PA_{fv} = \frac{(FA_n)(i)}{(1+i)^n - 1} \tag{2.33}$$

where

PA_{fv} is the value of annuity payments when the future value of the annuity is given.
FA_n is the future value of the annuity after n interest periods or years.
n is total number of interest periods or years.
i is annual compound interest rate.

Similarly, from Equation (2.32), the value of annuity payments when the present value of the annuity is given is expressed by

$$PA_{pv} = \frac{(PV)(i)}{1-(1+i)^{-n}} \tag{2.34}$$

where

PA_{pv} is the value of annuity payments when the present value of the annuity is known.

PV is present value of all payments.

Example 2.7

Assume that a firm plans to acquire a facility at the end of the next 5 years. The
estimated cost of the facility after the specified period is $800,000. The firm has
decided to make deposits of equal amounts of money at the end of each of next
5 years so that the total amount accumulates to $800,000. Calculate the amount
of money the firm should deposit at the end of each year, if the annual compound
interest rate is 8%.

By substituting the given data values into Equation (2.33), we get

$$PA_{fv} = \frac{(800,000)(0.08)}{(1+0.08)^5 - 1}$$

$$= \$136,365.16$$

This means that the firm should deposit $136,365.16 at the end of each year to
fulfill its objective.

Example 2.8

Assume that we have the following data values:

$PV = \$400,000$, $i = 4\%$, and $n = 7$ years

Using Equation (2.34), calculate the value of annuity payments.

Using the given data values in Equation (2.34) yields

$$PA_{pv} = \frac{(400,000)(0.04)}{1-(1+0.04)^{-7}}$$

$$= \$66,643.84$$

Thus, the value of annuity payments is $66,643.84.

2.6 Depreciation Methods

The term *depreciation* simply means decline in value. There are different types of depreciation with respect to engineering equipment: monetary depreciation, technological depreciation, physical depreciation, and

functional depreciation [3]. Over the years, a number of methods with respect to monetary depreciation have been developed. Three of these methods are presented next [2–4].

2.6.1 Sum-of-Years-Digits (SYD) Method

The name of this method is derived from the calculation procedure used. The method provides a larger depreciation charge during early life years of the equipment, system, or product than during its later life years.

The annual depreciation charge is expressed by [2,4]

$$DC_a = (C_a - V_s) \left[\frac{(L_S - n + 1)}{(1 + 2 + 3 + \cdots + L_S)} \right]$$

$$= (C_a - V_S)(2)(L_S - n + 1) / L_S(L_S + 1)$$

(2.35)

where

DC_a is annual depreciation charge.
C_a is product or item acquisition cost.
V_S is product or item salvage value at the end of its service life.
L_S is product or item service life expressed in years.
n is total number of years of the product or item in actual service.

The book value of the product or item at the end of year n is given by [4]

$$V_{bn} = 2(C_a - V_s) \left[\frac{1 + 2 + 3 + \cdots + (L_S - n)}{L_S(L_S + 1)} \right] + V_S$$

(2.36)

where V_{bn} is product or item book value at the end of year n.

Example 2.9

Assume that the cost, useful life, and salvage value after the useful life of an engineering system are $900,000, 10 years, and $60,000, respectively. Calculate the system book value at the end of year 5 by using the SYD method.

By substituting the given data values into Equation (2.36), we obtain

$$V_{b5} = 2 \, (900,000 - 60,000) \left[\frac{1 + 2 + 3 + \cdots + (10 - 5)}{10(10 + 1)} \right] + 60,000$$

$$= \$289,090.9$$

Thus, the system book value at the end of year 5 is $289,090.90.

2.6.2 Straight-Line Method

This method assumes the linear decrease with time in the value of an item, product, or system. Thus, during the service life of the item, product, or system an equal sum of money is charged each year for depreciation. The annual depreciation is expressed by

$$DC_a = (C_a - V_S)/L_S \tag{2.37}$$

The book value of the product, item, or system at the end of year n is given by

$$V_{bn} = C_a - n(DC_a) \tag{2.38}$$

Using Equation (2.37) in Equation (2.38) yields

$$V_{bn} = C_a - n\left[\frac{(C_a - V_S)}{L_S}\right] \tag{2.39}$$

Example 2.10

Assume that the acquisition cost, the expected useful life, and salvage value after the useful life of a piece of equipment are $600,000, 12 years, and $30,000, respectively. The equipment annual depreciation charge is constant. Calculate the equipment annual depreciation charge.

By substituting the given data values into Equation (2.37), we get

$$DC_a = (600,000 - 30,000)/12$$

$$= \$47,500$$

Thus, the equipment annual depreciation charge is $47,500.

2.6.3 Declining-Balance Method

This method is also known as the Matheson formula or the constant percentage method. In this approach, the annual depreciation is a fixed percentage of the book value at the beginning of the year. Although the annual depreciation is different for each year, the declining-balance (i.e., fixed-percentage) factor remains constant throughout the useful life of the equipment or item.

This method writes off the cost of the equipment or item early in its life at an accelerated rate and at correspondingly lower annual charges close to the final years of the equipment or item service. The depreciation factor or rate is expressed by

$$R_d = 1 - \left[\frac{V_S}{C_a} \right]^{1/L_S}$$ (2.40)

where R_d is the depreciation rate or factor. Note that this method assumes that the salvage value of the equipment or item is always positive.

The book value of the equipment or item at the end of year n is defined by

$$V_{bn} = C_a (1 - R_d)^n$$ (2.41)

By inserting Equation (2.40) into Equation (2.41), we get

$$V_{bn} = C_a \left[\frac{V_S}{C_a} \right]^{n/L_S}$$ (2.42)

The annual depreciation charge is defined by

$$DC_a = [V_{b(n-1)}][R_d]$$ (2.43)

where $V_{b(n-1)}$ is the equipment or item book value at $(n-1)$ years.

Using Equation (2.40) in Equation (2.43) yields

$$DC_a = [V_{b(n-1)}] \left[1 - \left\{ \frac{V_S}{C_a} \right\}^{1/L_S} \right]$$ (2.44)

Example 2.11

Assume that the cost, useful life, and salvage value after the useful life of a piece of engineering equipment are $700,000, 15 years, and $80,000, respectively. Calculate the equipment book value at the end of year 10 by using the declining-balance method.

By substituting the specified data values into Equation (2.42), we obtain

$$V_{b10} = (700,000) \left[\frac{80,000}{700,000} \right]^{10/15}$$

$$= \$164,851.4$$

Thus, the equipment book value at the end of year 10 is $164,851.40.

Problems

1. What is the difference between simple interest and compound interest?
2. Define the following terms:
   - present value
   - future amount
   - depreciation
3. A company borrowed $400,000 for a period of 5 years at an annual simple interest rate of 6% to procure an engineering system. Calculate the total amount of money the company has to pay to the lender at the end of 5 years.
4. Prove the following equation:

$$A_n = P(1+i)^n \qquad (2.45)$$

   where
   > n is the number of interest periods.
   > A_n is the total or compound amount at the end of the nth interest period.
   > P is the principal amount (i.e., borrowed).

5. What is effective annual interest rate?
6. An individual deposited $90,000 in a bank at a nominal interest rate of 7% compounded monthly, for 10 months. Estimate the value of the effective annual interest rate.
7. Assume that the total operation and maintenance cost of an engineering system after its 7-year usage will be $100,000. Calculate the present value of $100,000 if the annual compound interest rate is 4%.
8. A company plans to procure a facility at the end of the next 7 years. The estimated cost of the facility after the specified period is $1,000,000. The company has decided to make deposits of equal sums of money at the end of each of the next 7 years so that the total amount accumulates to $1,000,000. Calculate the amount of money the company should deposit at the end of each year, if the annual compound interest is 6%.
9. Assume that the cost, useful life, and salvage value after the useful life of a piece of engineering equipment are $660,000, 8 years, and $40,000, respectively. The equipment annual depreciation charge is constant. Calculate the equipment annual depreciation charge by using the straight-line method.
10. Compare the SYD and declining-balance depreciation methods.

References

1. Paul-DeGarmo, E., Canada, J. R., and Sullivan, W. G. 1979. *Engineering economy.* New York: Macmillan Publishing Company, Inc.
2. Dhillon, B. S. 1989. *Life cycle costing: Techniques, models, and applications.* New York: Gordon and Breach Science Publishers.
3. Riggs, J. L. 1981. *Production systems: Planning, analysis, and control.* New York: John Wiley & Sons.
4. Riggs, J. L. 1968. *Economic decision models for engineers and managers.* New York: McGraw–Hill Book Company.

3

Life Cycle Costing Fundamentals

3.1 Introduction

Past experience indicates that engineering equipment procured at the lowest cost may not necessarily be that which also costs the least amount of money over its useful life. More specifically, the equipment ownership cost could be quite significant and frequently exceeds the procurement cost. For example, various studies performed by the U.S. Department of Defense indicate that the maintenance cost over equipment's useful life could be many times the procurement cost [1,2].

In fact, by simply examining the Defense Department's overall annual budget, it can easily be observed that operation and maintenance costs are an important factor. For example, in fiscal year 1974, 27% of the overall budget of the Department of Defense accounted for operation and maintenance activities and 20% was for procurement [3,4]. This simply means that, in equipment acquisition analysis, it is important to consider the cost of equipment ownership. Otherwise, procurement decisions based totally on the acquisition cost may not be the best decision in the long term.

The approach used for estimating the total life cycle cost of equipment procurement is known as life cycle costing. This chapter presents various fundamental aspects of this approach.

3.2 Need and Information Required for Life Cycle Costing

Life cycle costing is increasingly being used in the industrial sector around the world to make various types of decisions that directly or indirectly concern engineering equipment and systems. There could be many reasons for this upward trend, such as [4]

- competition;
- increasing operation and maintenance costs;
- budget limitations;
- expensive products or systems (e.g., military systems, space systems, and aircraft);

- rising inflation; and
- increasing awareness of cost effectiveness among product, equipment, and system users.

Various types of information are required to perform life cycle costing studies. These include the acquisition cost of the item, the useful operational life of the item in years, the annual maintenance cost of the item, transportation (delivery) and installation costs of the item, discount and escalation rates, the annual operating cost of the item, taxes (e.g., tax benefits from depreciation, investment tax credit), and the salvage value or disposal cost of the item [5].

In any case, prior to starting a life cycle costing study, it is considered useful to seek answers to questions on topics such as the following [6,7]:

- goal of the estimate;
- assumptions and ground rules;
- treatment of uncertainties;
- required data;
- required details of the analysis and analysis-related constraints;
- involved personnel and the responsibility of the cost analyst;
- controlling and auditing the life cycle costing process by the seller's and purchaser's management;
- estimating procedures to be followed;
- life cycle cost analysis users;
- life cycle cost analysis format;
- life cycle costing time schedule;
- required accuracy and precision of the analysis; and
- fund limitations.

3.3 Life Cycle Costing Application Areas

Life cycle costing can be used in a large number of areas. The six primary uses of life cycle cost include [6]:

- selecting among competing bidders for a project;
- long-range planning and budgeting;
- controlling an ongoing project;
- comparing competing projects;

- deciding the replacement of aging equipment; and
- comparing logistics concepts.

Lamar [8] has presented the following somewhat more specific applications of life cycle cost analyses:

- determining cost drivers;
- forecasting future budget needs;
- selecting the most effective procurement strategy;
- improving comprehension of fundamental design-related parameters in equipment or system product design and development;
- formulating contractor incentives;
- making strategic decisions and design trade-offs;
- optimizing appropriate training needs;
- choosing among options;
- providing effective objectives for program control;
- assessing new technology application; and
- carrying out source selections.

3.4 Life Cycle Costing Activities and Associated Steps

Many activities are associated with life cycle costing. Some of these include [9]:

- defining an item's or a product's life cycle;
- identifying all cost drivers;
- establishing escalated and discounted life cycle costs;
- developing an accounting breakdown structure;
- establishing cost estimating relationships for each and every component in the life cycle cost breakdown structure;
- developing constant dollar cost profiles;
- defining activities that generate an item's or a product's ownership costs;
- conducting appropriate sensitivity analysis; and
- identifying cause and effect relationships.

Over the years, many authors have proposed steps for performing life cycle cost analysis [10–13]. Figure 3.1 shows 10 steps considered quite effective in

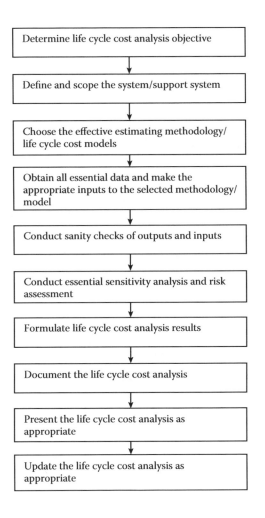

FIGURE 3.1
Steps for performing life cycle cost analysis.

performing life cycle cost analysis [14]. Additional information on these steps is available in reference 14.

3.5 Approach for Incorporating Life Cycle Costing into the Planning Process for Proposals and Contracts

Over the years, equipment or system procurement contracts requiring contractor or manufacturer commitments for equipment or system life cycle cost have increased quite significantly. Many of these contractors and manufacturers are not familiar with life cycle cost-related acquisitions. In order to overcome this shortcoming, a six-step approach for these contractors and manufacturers to prepare for life cycle cost-related acquisitions follows [15]:

- *Organize for life cycle costing.* This step is basically concerned with establishing a proper organization for life cycle costing and assigning life cycle cost responsibilities.
- *Gather and develop background information related to life cycle costing.* This step calls for becoming acquainted with the existing life cycle cost estimation models and components of the life cycle cost considered vital to the company's product and equipment.
- *Perform analysis of all requirements for life cycle costing–related response.* This step involves tasks such as performing analysis of likely life cycle cost estimation model components to determine the types of data required for life cycle cost response and performing analysis of the information considered essential for management decision making.
- *Develop a plan for the life cycle costing technical proposal.* This step is basically concerned with planning the life cycle costing–related response for a technical proposal under consideration.
- *Develop a plan to identify and analyze life cycle cost risk.* This step calls for developing a plan to identify risk areas and address methods to analyze such risks when life cycle cost–related guarantees are committed as an element of a proposed procurement.
- *Develop a plan to achieve life cycle cost goals.* This step involves developing a plan to achieve the set life cycle cost goals during the specified contract period.

3.6 Areas for Evaluating a Life Cycle Costing Program

In order to keep a life cycle costing program in good order, it is essential to evaluate it periodically. There are many areas in which questions could be raised to determine the effectiveness of the life cycle costing program. Some of these areas include [4,6,16]:

- effectiveness of cost-estimating techniques used;
- cost model construction;
- broadness of cost-estimating database;
- identification of all cost drivers;
- proper consideration of discounting and inflation factors;

- performance of trade-off studies;
- inclusion of all life cycle costing–related requirements into design subcontracts;
- cost performance review of subcontractors;
- cost estimates' validation through an independent appraisal;
- life cycle costing management representative's qualifications;
- coordination of life cycle cost and design to cost-related activities;
- defining of cost priority with respect to factors such as product performance, delivery schedule, and other requirements by management;
- formal notifications to all organizations or departments involved in the life cycle costing program regarding their cost goals;
- compatibility of system safety, reliability, and maintainability programs with life cycle cost–related requirements; and
- awareness of the buyer regarding the top 10 cost drivers and proper suggestions to reduce such costs.

3.7 Life Cycle Costing Data Sources

In order to perform effective life cycle cost analysis, the availability of reliable cost data is vital. This means that the existence of good cost data banks is very important. Thus, in developing a new cost data bank, careful attention must be given to factors such as comprehensiveness, size, uniformity, flexibility, responsiveness, ready accessibility, orientation, and expansion or contraction capability [17]. Furthermore, at a minimum, a life cycle costing data bank should incorporate information such as user pattern records, descriptive records (hardware and site), cost records, and procedural records (operation and maintenance).

Although data for life cycle cost analysis can be obtained from many sources, their amount and quality may vary quite considerably. Therefore, prior to starting a life cycle cost study, it is important to examine carefully factors such as data bias, data applicability, data availability, data comparability to other existing data, data orientation toward the problem under consideration, and data coordination with other information. Some of the important sources for obtaining cost-related data include [4,17,18]:

- costs for pressure vessels [19];
- American Building Owners and Managers Association (BOMA) handbook;
- costs for solid waste shredders [20];

- costs for heat exchangers [21–23];
- unit price manuals: Marshall and Swift, means, Dodge, Richardson, and building cost file;
- cost analysis cost estimating (CACE) model [24,25];
- costs for varieties of process equipment [26–29];
- budgeting annual cost estimating (BACE) model [24,25];
- programmed review of information for costing and evaluation (PRICE) model [24]; and
- costs for motors, storage tanks, centrifugal pumps, etc. [30,31].

3.8 Life Cycle Costing Advantages and Disadvantages and Related Important Points

Over the years, various advantages and disadvantages of life cycle costing have been identified by various professionals. Some of the important advantages of life cycle costing are shown in Figure 3.2 [4]. In contrast, some of the main disadvantages of life cycle costing include that it

- is time consuming;
- is costly;
- has doubtful data accuracy; and
- is a trying task when attempting to obtain data for analysis.

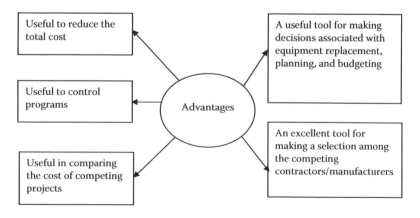

FIGURE 3.2
Life cycle costing advantages.

Many important points are associated with life cycle costing, some of which include:

- The main goal of life cycle costing is to get the maximum benefit from limited resources.
- Management plays a key role in making life cycle costing a worthwhile effort.
- Risk management is the essence of life cycle costing in general.
- The availability of good data is very important for good life cycle cost estimates.
- The life cycle cost model must include all program-related costs.
- There is a definite need for both the product manufacturer and the user to organize effectively to control life cycle cost.
- There is a definite need to perform trade-offs among life cycle cost, design to cost, and performance throughout the life of the program.
- Some surprises may still occur, even when the estimator is very competent.
- Life cycle costing is gaining importance as a method for performing design optimization, making strategic decisions, conducting detailed trade-off studies, etc.
- A highly knowledgeable and experienced cost analyst may compensate for various database-related difficulties.

3.9 Life Cycle Costing Concept Application in Selecting Equipment from Competing Manufacturers

From time to time, equipment or system users are faced with selecting the most cost-effective equipment or system from a number of competing manufacturers. In situations such as these, life cycle costing becomes a useful tool. The application of the life cycle costing concept in selecting the most cost-effective equipment from competing manufacturers is demonstrated through Example 3.1.

Example 3.1

A company using machining equipment to manufacture a certain type of engineering part is contemplating replacing it with a better version. Four different pieces of machining equipment, manufactured by four different manufacturers, are being considered for its replacement; their data are presented in Table 3.1.

TABLE 3.1

Data for Four Types of Machining Equipment under Consideration

No.	Description	Machining Equipment A	Machining Equipment B	Machining Equipment C	Machining Equipment D
1	Procurement cost	$300,000	$270,000	$290,000	$350,000
2	Expected useful life in years	10	10	10	10
3	Annual failure rate	0.08	0.07	0.06	0.04
4	Cost of a failure	$2,000	$2,500	$3,000	$1,000
5	Annual interest rate	6%	6%	6%	6%
6	Annual operating cost	$6,000	$7,000	$6,500	$8,000

Determine which of the four pieces of machining equipment should be procured to replace the existing one in regard to their life cycle costs.

Life Cycle Cost Analysis: Machining Equipment A

The expected cost, C_{fa}, of failure per year of machining equipment A is given by

$$C_{fa} = (2,000)(0.08)$$

$$= \$160$$

where C_{fa} is the machining equipment A annual expected failure cost.

Using Chapter 2 and reference 4, the present value, PV_{af}, of machining equipment A life cycle failure cost is expressed by

$$PV_{af} = C_{fa}\left[\frac{1-(1+i)^{-k}}{i}\right] \tag{3.1}$$

where

PV_{af} is present value of machining equipment A life cycle failure cost.

i is annual interest rate.

k is machining equipment's expected useful life in years.

By substituting the preceding calculated value and the given data values into Equation (3.1), we get

$$PV_{af} = (160)\left[\frac{1-(1+0.06)^{-10}}{0.06}\right]$$

$$= \$1176.61$$

Similarly, using Chapter 2 and reference 4, the present value, PV_{ao}, of machining equipment A life cycle operating cost is given by

$$PV_{ao} = C_{oa}\left[\frac{1-(1+i)^{-k}}{i}\right] \tag{3.2}$$

where
PV_{ao} is present value of machining equipment A life cycle operating cost.
C_{oa} is machining equipment A annual operating cost.

By substituting the given data values into Equation (3.2), we get

$$PV_{ao} = (6,000)\left[\frac{1-(1+0.06)^{-10}}{0.06}\right]$$

$$= \$44,160.52$$

Thus, the life cycle cost of machining equipment A is given by

$$LCC_a = PC_a + PV_{af} + PV_{ao} \tag{3.3}$$

where
LCC_a is machining equipment A life cycle cost.
PC_a is machining equipment A procurement cost.

By substituting the given data value and the preceding calculated values into Equation (3.3), we obtain

$$LCC_a = 300,000 + 1176.61 + 44,160.52$$

$$= \$345,337.13$$

Life Cycle Cost Analysis: Machining Equipment B

The expected cost, C_{fb}, of failure per year of machining equipment B is given by

$$C_{fb} = (2,500)(0.07)$$

$$= \$175$$

where C_{fb} is machining equipment B annual expected failure cost.
Using Chapter 2 and reference 4, the present value, PV_{bf}, of machining equipment B life cycle failure cost is given by

$$PV_{bf} = C_{fb}\left[\frac{1-(1+i)^{-k}}{i}\right] \tag{3.4}$$

where PV_{bf} is present value of machining equipment B life cycle failure cost.

By substituting the preceding calculated value and the given data values into Equation (3.4), we get

$$PV_{bf} = (175)\left[\frac{1-(1+0.06)^{-10}}{0.06}\right]$$

$$= \$1,288.01$$

Similarly, using Chapter 2 and reference 4, the present value, PV_{bo}, of machining equipment B life cycle operating cost is expressed by

$$PV_{bo} = C_{ob}\left[\frac{1-(1+i)^{-k}}{i}\right] \quad (3.5)$$

where
PV_{bo} is present value of machining equipment B life cycle operating cost.
C_{ob} is machining equipment B annual operating cost.

By substituting the given data values into Equation (3.5), we obtain

$$PV_{bo} = (7,000)\left[\frac{1-(1+0.06)^{-10}}{0.06}\right]$$

$$= \$51,520.61$$

Thus, the life cycle cost of machining equipment B is given by

$$LCC_b = PC_b + PV_{bf} + PV_{bo} \quad (3.6)$$

where
LCC_b is machining equipment B life cycle cost.
PC_b is machining equipment B procurement cost.

By substituting the given data value and the preceding calculated values into Equation (3.6), we get

$$LCC_b = 270,000 + 1,288.01 + 51,520.61$$

$$= \$322,808.62$$

Life Cycle Cost Analysis: Machining Equipment C

The expected cost, C_{fc}, of failure per year of machining equipment C is given by

$$C_{fc} = (3,000)(0.06)$$

$$= \$180$$

where C_{fc} is machining equipment C annual expected failure cost.

Using Chapter 2 and reference 4, the present value, PV_{cf}, of machining equipment C life cycle failure cost is expressed by

$$PV_{cf} = C_{fc}\left[\frac{1-(1+i)^{-k}}{i}\right]$$

(3.7)

where PV_{cf} is present value of machining equipment C life cycle failure cost.

By substituting the preceding calculated value and the given data values into Equation (3.7), we get

$$PV_{cf} = (180)\left[\frac{1-(1+0.06)^{-10}}{0.06}\right]$$

$$= \$1,324.81$$

Similarly, using Chapter 2 and reference 4, the present value, PV_{co}, of machining equipment C life cycle operating cost is expressed by

$$PV_{co} = C_{oc}\left[\frac{1-(1+i)^{-k}}{i}\right]$$

(3.8)

where

PV_{co} is present value of machining equipment C life cycle operating cost.
C_{oc} is machining equipment C annual operating cost.

By substituting the given data values into Equation (3.8), we get

$$PV_{co} = (6,500)\left[\frac{1-(1+0.06)^{-10}}{0.06}\right]$$

$$= \$47,840.56$$

Thus, the life cycle cost of machining equipment C is given by

$$LCC_c = PC_c + PV_{cf} + PV_{co}$$

(3.9)

where

LCC_c is machining equipment C life cycle cost.
PC_c is machining equipment C procurement cost.

By substituting the given data value and the preceding calculated values into Equation (3.9), we get

$$LCC_c = 290,000 + 1,324.81 + 47,840.56$$

$$= \$339,165.37$$

Life Cycle Cost Analysis: Machining Equipment D

The expected cost, C_{fd}, of failure per year of machining equipment D is given by

$$C_{fd} = (1,000)(0.04)$$
$$= \$40$$

where C_{fd} is machining equipment D annual expected failure cost.

Using Chapter 2 and reference 4, the present value, PV_{df}, of machining equipment D life cycle failure cost is expressed by

$$PV_{df} = C_{fd} \left[\frac{1-(1+i)^{-k}}{i} \right] \tag{3.10}$$

where PV_{df} is present value of machining equipment D life cycle failure cost.

By substituting the preceding calculated value and the given data values into Equation (3.10), we get

$$PV_{df} = (40) \left[\frac{1-(1+0.06)^{-10}}{0.06} \right]$$
$$= \$294.40$$

Similarly, using Chapter 2 and reference 4, the present value, PV_{do}, of machining equipment D life cycle operating cost is given by

$$PV_{do} = C_{od} \left[\frac{1-(1+i)^{-k}}{i} \right] \tag{3.11}$$

where
PV_{do} is present value of machining equipment D life cycle operating cost.
C_{od} is machining equipment D annual operating cost.

By substituting the given data values into Equation (3.11), we obtain

$$PV_{do} = (8,000) \left[\frac{1-(1+0.06)^{-10}}{0.06} \right]$$
$$= \$58,880.69$$

Thus, the life cycle cost of machining equipment D is expressed by

$$LCC_d = PC_d + PV_{df} + PV_{do} \tag{3.12}$$

where
LCC_d is machining equipment D life cycle cost.
PC_d is machining equipment D procurement cost.

By substituting the given data value and the preceding calculated values into Equation (3.12), we get

$$LCC_d = 350,000 + 294.40 + 58,880.69$$

$$= \$409,175.09$$

Thus, the life cycle costs of machining equipment A, B, C, and D are \$345,337.13, \$322,808.62, \$339,165.37, and \$409,175.09, respectively. By examining these values, it is concluded that machining equipment B should be purchased because its life cycle cost is the lowest.

Problems

1. Write an essay on life cycle costing fundamentals.
2. Discuss the need for life cycle costing.
3. List at least 10 specific applications of life cycle cost analyses.
4. List at least eight activities associated with life cycle costing.
5. What are the steps used to perform life cycle cost analysis?
6. Describe the six-step approach for unfamiliar contractors and manufacturers to prepare for life cycle cost–related acquisitions.
7. List at least 12 areas on which questions could be raised to determine the effectiveness of a life cycle costing program.
8. List at least 10 important sources for obtaining cost-related data.
9. What are the advantages and disadvantages of life cycle costing?
10. A company using a machine to manufacture a certain type of engineering part is contemplating replacing it with a better one. Two different machines are being considered for its replacement and their data are presented in Table 3.2. Determine which of the two machines should be procured to replace the existing machine in regard to their life cycle costs.

TABLE 3.2

Data for Two Machines under Consideration

No.	Description	Machine A	Machine B
1	Procurement cost	\$140,000	\$170,000
2	Annual failure rate	0.03	0.04
3	Expected useful life in years	12	12
4	Annual operating cost	\$6,000	\$4,000
5	Cost of a failure	\$12,000	\$13,000
6	Annual interest rate	8%	8%

References

1. Ryan, W. J. 1968. Procurement views of life cycle costing. *Proceedings of the Annual Symposium on Reliability* 164–168.
2. Dhillon, B. S. 1983. *Reliability engineering in systems design and operations.* New York: Van Nostrand Reinhold Company.
3. Louis-Wienecke, E., and Feltus, E. E. 1979. Predictive operations and maintenance cost model. Report no. ADA078052. Available from the National Technical Information Service (NTIS), Springfield, VA.
4. Dhillon, B. S. 1989. *Life cycle costing: Techniques, models, and applications.* New York: Gordon and Breach Science Publishers.
5. Brown, R. J. 1979. A new marketing tool: Life cycle costing. *Industrial Marketing Management* 8:109–113.
6. Robert-Seldon, M. 1979. *Life cycle costing: A better method of government procurement.* Boulder, CO: Westview Press.
7. Reiche, H. 1980. Life cycle cost. In *Reliability and maintainability of electronic systems,* ed. J. E. Arsenault and J. A. Roberts, 3–23. Potomac, MD: Computer Science Press.
8. Lamar, W. E. 1981. Technical evaluation report on design to cost and life cycle cost. North Atlantic Treaty Organization Advisory Group for Aerospace Research and Development (AGARD) advisory report no. 165. Available from the National Technical Information Service (NTIS), Springfield, VA.
9. Earles, M. 1981. *Factors, formulas, and structures for life cycle costing.* Concord, MA: Eddins–Earles.
10. Kaufman, R. J. 1969. Life cycle costing: Decision making tool for capital equipment acquisitions. *Journal of Purchasing* 5:16–31.
11. Kaufman, R. J. 1969. Life cycle costing: For capital equipment decisions. *Automation* March: 75–80.
12. Coe, C. K. 1981. Life cycle costing by state governments. *Public Administration Review* September/October: 564–569.
13. Wynholds, H. W., and Skratt, J. P. 1977. Weapon system parametric life cycle cost analysis. *Proceedings of the Annual Reliability and Maintainability Symposium* 303–309.
14. Greene, L. E., and Shaw, B. L. 1990. The steps for successful life cycle cost analysis. *Proceedings of the IEEE National Aerospace and Electronics Conference* 1209–1216.
15. Schmidt, B. A. 1979. Preparation for LCC proposals and contracts. *Proceedings of the Annual Reliability and Maintainability Symposium* 62–66.
16. Bidwell, R. L. 1977. Checklist for evaluating LCC program effectiveness. Product Engineering Services Office, U.S. Department of Defense, Washington, D.C.
17. Bowen, B., and Williams, J. 1975. Life costing and problems of data. *Industrialization Forum* 6:21–24.
18. Dhillon, B. S. 1999. *Design reliability: Fundamentals and applications.* Boca Raton, FL: CRC Press.
19. Mulet, A., Corripio, A. B., and Evans, L. B. 1981. Estimate cost of pressure vessels via correlations. *Chemical Engineering* 88:20, 456.
20. Fang, C. S. 1980. The cost of shredding municipal solid waste. *Chemical Engineering* 87 (7): 151–153.

21. Purohit, G. P. 1985. Cost of double pipe and multi-tube heat exchangers. *Chemical Engineering* 92:96–97.
22. Woods, D. R., Anderson, S. J., and Norman, S. L. 1976. Evaluation of capital cost data: Heat exchangers. *Canadian Journal of Chemical Engineering* 54:469.
23. Kumana, J. D. 1984. Cost update on specialty heat exchangers. *Chemical Engineering* 91 (13): 164.
24. Marks, K. E., Garrison-Massey, H., and Bradley, B. D. 1978. An appraisal of models used in life cycle cost-estimation for U.S. Air Force (USAF) aircraft systems. Report no. R-2287-AF. Prepared by the Rand Corporation, Santa Monica, CA.
25. Department of the Air Force. 1975. USAF cost and planning factors. Report no. AFR 173-10, Washington, D.C.
26. Hall, R. S., Mately, J., and McNaughton, K. J. 1982. Current costs of process equipment. *Chemical Engineering* 87 (7): 80.
27. Klumpar, I. V., and Slavsky, S. T. 1985. Updated cost factors: Process equipment, commodity materials, and installation labor. *Chemical Engineering* 92 (15): 73–74.
28. Humphreys, K. K., and Katell, S. 1981. *Basic cost engineering.* New York: Marcel Dekker.
29. Peters, M. S., and Timmerhaus, K. D. 1980. *Plant design and economics for chemical engineers.* New York: McGraw–Hill Book Company.
30. Corripio, A. B., Chrien, K. S., and Evans, L. B. 1982. Estimate costs of heat exchangers and storage tanks via correlations. *Chemical Engineering* 89 (2): 125–126.
31. Corripio, A. B., Chrien, K. S., and Evans, L. B. 1982. Estimate costs of centrifugal pumps and electric motors. *Chemical Engineering* 89 (4): 115.

4

Life Cycle Cost Models and
Cost Estimation Methods

4.1 Introduction

Over the years, a large number of life cycle cost models have been developed that include both general and specific models [1,2]. No single life cycle cost model has been accepted as a standard model in the industrial sector. There could be many reasons for not having a standard model, including the inclinations of users, the nature of the problem, the existence of many different cost data collection systems, and many different types of equipment, devices, or systems. Nonetheless, irrespective of the types of models used in performing life cycle cost analysis, they all must be effective in representing equipment, systems, or subsystems, transparent and visible.

Cost estimating is an important activity because estimated cost has to be as close as possible to actual value; otherwise, an incorrect estimate may lead to serious consequences of various types. Currently, many methods are used to estimate various types of costs. Each one has its advantages and disadvantages. More specifically, a cost estimation method or approach may be very effective in one type of application and rather weak in another. This simply means that utmost care is necessary in selecting a cost estimation method for a specific application.

This chapter presents some of the life cycle cost models and cost estimation methods considered useful in performing life cycle cost analysis.

4.2 Types of Life Cycle Cost Models and Their Inputs

Over the years, life cycle cost models have been classified under various categories [3–6]. For example, Gupta [3] and Sherif and Kolarik [5] have classified life cycle cost models under three categories: conceptual models, analytical models, and heuristic models. The conceptual models are quite flexible but have rather limited applications; they are usually based on the hypothesized relationships of variables given in a qualitative fashion. One example of the conceptual models is available in Goldman and Slattery [7].

The analytical models are based on some sort of mathematical relationship and their subcategories include logistic support models, design trade models, and the total cost models. Finally, the heuristic models may be described simply as the ill-structured version of the analytical models. An example of these models is available in Kolarik [8]. Overall, in this chapter, the life cycle cost models are simply classified under two categories: general life cycle cost models and specific life cycle cost models.

There are many inputs to life cycle cost models. These include [6,9]:

- warranty coverage period;
- average material cost of a failure;
- cost of training;
- cost of installation;
- system's or item's listed price;
- cost of carrying spares in inventory;
- mean time between failures;
- mean time to repair;
- spares' requirements;
- cost of labor per corrective maintenance action; and
- time spent for travel.

4.3 General Life Cycle Cost Models

The general life cycle cost models are not tied to any specific system or equipment. Some of these models are presented next.

4.3.1 General Life Cycle Cost Model I

In this case, the equipment or system life cycle cost is divided into two main parts: recurring cost and nonrecurring cost. Thus, the system or equipment life cycle cost is expressed by [10]

$$LCC = RC + NRC \qquad (4.1)$$

where
LCC is item or system life cycle cost.
RC is recurring cost.
NRC is nonrecurring cost.

The recurring cost, *RC*, is expressed by

$$RC = OC + IC + SC + MC + MTC \qquad (4.2)$$

where
OC is operating cost.
IC is inventory cost.
SC is support cost.
MC is manpower cost.
MTC is maintenance cost.

The nonrecurring cost, *NRC*, is expressed by

$$NRC = C_p + C_i + C_q + C_r + C_t + C_{rm} + C_s \qquad (4.3)$$

where
C_p is procurement cost.
C_i is installation cost.
C_q is qualification approval cost.
C_r is research and development cost.
C_t is training cost.
C_{rm} is reliability and maintainability improvement cost.
C_s is support cost.

4.3.2 General Life Cycle Cost Model II

In this case, the equipment or system life cycle cost is divided into three main parts: procurement cost, initial logistic cost, and recurring cost. Thus, the system or equipment life cycle cost is expressed by [11]

$$LCC = C_1 + C_2 + C_3 \qquad (4.4)$$

where
LCC is item or system life cycle cost.
C_1 is acquisition or procurement cost.
C_2 is initial logistic cost.
C_3 is recurring cost.

The initial logistic cost, C_2, is composed of one-time costs such as the cost of procurement of new support equipment not accounted for in the life cycle costing solicitation and training, the cost of existing support equipment modifications, and the cost of initial technical data management.

The three main components of the recurring cost, C_3, are operating cost, management cost, and maintenance cost.

4.3.3 General Life Cycle Cost Model III

This model was developed by the U.S. Navy to estimate life cycle cost of major weapon systems [12,13]. The system life cycle cost is divided into five main parts: research and development cost, the cost of associated systems, investment cost, termination cost, and operating and support cost. Thus, the system life cycle cost is expressed by

$$LCC = C_1 + C_2 + C_3 + C_4 + C_5 \tag{4.5}$$

where
LCC is system life cycle cost.
C_1 is research and development cost.
C_2 is cost of associated systems.
C_3 is investment cost.
C_4 is termination cost.
C_5 is operating and support cost.

The two main components of the research and development cost, C_1, are full-scale development cost and validation cost. Similarly, the two main elements of the cost of associated systems, C_2, are their investment cost and their operating and support cost.

The investment cost, C_3, is also made up of two main components: the government investment cost and the procurement cost. The termination cost, C_4, is expressed by

$$C_4 = \sum_{i=1}^{m} x_i \, c_t \tag{4.6}$$

where
m is total number of years in the life cycle.
x_i is total number of major system items put out of action during year i.
c_t is terminal cost of the major system item.

Finally, the elements of the operating and support cost are depot supply cost, depot maintenance cost, operating cost, personnel support and training costs, sustaining investment cost, installation support cost, second destination transportation cost, and organizational and intermediate maintenance activity cost.

4.3.4 General Life Cycle Cost Model IV

In this case, the life cycle cost is expressed by [2,14]

$$LCC = C_{cp} + C_{dp} + C_{pp} + C_{op} \tag{4.7}$$

where

   LCC is life cycle cost.
   C_{cp} is cost associated with the conceptual phase.
   C_{dp} is cost associated with the definition phase.
   C_{pp} is cost associated with the procurement phase.
   C_{op} is cost associated with the operational phase.

The costs of conceptual and definition phases are relatively small in comparison to the costs of procurement and operational phases. They are basically associated with the labor effort.

The four main elements of the procurement phase cost are the cost of the prime equipment or system, the cost of acquisition personnel, the cost of support equipment, and the cost of program management. Finally, the operational phase cost is expressed by

$$C_{op} = C_m + C_{fo} + C_{oa} \tag{4.8}$$

where

   C_m is maintenance cost.
   C_{fo} is functional operating cost.
   C_{oa} is operational administrative cost.

Additional information on this model is available in Dhillon [2] and Stordahl and Short [14].

4.3.5 General Life Cycle Cost Model V

In this case, the life cycle cost is expressed by [6,15]

$$LCC = C_{rd} + C_{pc} + C_{os} + C_{rt} \tag{4.9}$$

where

   LCC is life cycle cost.
   C_{rd} is research and development cost.
   C_{pc} is production and construction cost.
   C_{os} is operation and support cost.
   C_{rt} is retirement and disposal cost.

The research and development, C_{rd}, is expressed by

$$C_{rd} = \sum_{j=1}^{7} C_{rdj} \qquad (4.10)$$

where C_{rdj} is the jth cost element of the research and development cost for
  $j = 1$ (means product planning);
  $j = 2$ (means engineering design);
  $j = 3$ (means product or system life cycle management);
  $j = 4$ (means system or product test and evaluation);
  $j = 5$ (means product or system research);
  $j = 6$ (means product or system software); and
  $j = 7$ (means design documentation).

The production and construction cost, C_{pc}, is defined by

$$C_{pc} = \sum_{j=1}^{5} C_{pcj} \qquad (4.11)$$

where C_{pcj} is the jth cost element of the production and construction cost for
  $j = 1$ (means manufacturing);
  $j = 2$ (means construction);
  $j = 3$ (means quality control);
  $j = 4$ (means initial logistics support); and
  $j = 5$ (means industrial engineering and operations analysis).

The operation and support cost, C_{os}, is expressed by

$$C_{os} = \sum_{j=1}^{3} C_{osj} \qquad (4.12)$$

where C_{osj} is the jth cost element of the operation and support cost for
  $j = 1$ (means system or product operations);
  $j = 2$ (means product or system distribution); and
  $j = 3$ (means sustaining logistic support).

The retirement and disposal cost, C_{rt}, is defined by

$$C_{rt} = C_{ur} + [\theta K (C_{id} - r_v)] \qquad (4.13)$$

where

C_{ur} is ultimate retirement cost of the system or product.
θ is the condemnation factor.
K is total number of unscheduled maintenance actions.
C_{id} is item disposal cost.
r_v is reclamation value.

4.3.6 General Life Cycle Cost Model VI

This model was developed by the Material Command of the U.S. Army and is composed of three main components: investment cost, research and development cost, and operating and support cost [16–18]. Thus, the life cycle cost is expressed mathematically by [6,16–18]

$$LCC = C_1 + C_2 + C_3 \qquad (4.14)$$

where

LCC is life cycle cost.
C_1 is research and development cost.
C_2 is investment cost.
C_3 is operating and support cost.

The research and development cost, C_1, is composed of the following 10 components:

- research and development data cost;
- cost of research and development tooling;
- cost of research and development facilities;
- development engineering cost;
- prototype manufacturing cost;
- research and development test and evaluation cost;
- producibility engineering and planning cost;
- research and development system or project management cost;
- research and development training services and equipment cost; and
- other research and development costs.

The investment cost, C_2, is composed of 11 components:

- cost of production;
- initial training cost;

- transportation cost;
- cost of data;
- cost of engineering changes;
- nonrecurring investment cost;
- cost of system test and evaluation;
- production phase system or project management cost;
- cost of initial spares and repair parts;
- operational or site activation cost; and
- other investment costs.

Finally, the operating and support cost, C_3, is composed of six major components:

- cost of indirect support operations;
- cost of depot maintenance;
- cost of material modifications;
- consumption cost;
- cost of military personnel; and
- cost of other direct support operations.

Additional information on this model is available in references 16–18.

4.4 Specific Life Cycle Cost Models

Over the years, many mathematical models have been developed to estimate life cycle cost of specific systems or items. Some of these models are presented next.

4.4.1 Specific Life Cycle Cost Model I

This model is concerned with estimating the life cycle cost of switching power supplies, which is expressed by [19]

$$LCC_s = IC + FC \qquad (4.15)$$

where
  LCC_s is life cycle cost of switching power supplies.
  IC is initial cost.
  FC is failure cost.

The failure cost, *FC*, is expressed by

$$FC = \lambda(n)(C_r + C_s) \tag{4.16}$$

where
   λ is unit constant failure rate.
   n is expected life of the product/unit.
   C_r is repair cost.
   C_s is cost of spares.

The cost of spares, C_s, is defined by

$$C_s = C_u(K) \tag{4.17}$$

where
   C_u is unit spare cost.
   K is fractional number of spares for each active unit.

4.4.2 Specific Life Cycle Cost Model II

This model is concerned with estimating the life cycle cost of health care facilities. The health care facility life cycle cost is expressed by [6,13]

$$LCC_h = C_c + C_o \tag{4.18}$$

where
   LCC_h is health care facility life cycle cost.
   C_c is capital cost.
   C_o is operating cost.

The capital cost, C_o, is composed of the following eight cost components:

- land acquisition cost;
- financing cost;
- collateral equipment cost;
- direct construction or purchase cost;
- indirect cost;
- demolition and site preparation cost;
- alteration and replacement cost; and
- denial of use cost.

Similarly, the operating cost, C_o, is composed of the following 19 cost components:

- utilities and fuel cost;
- structural maintenance cost;
- heating system operations and maintenance cost;
- painting cost;
- equipment (furnishings) maintenance cost;
- exterior building cleaning cost;
- electrical system operations and maintenance cost;
- space changes cost;
- exterior restoration cost;
- grounds and roads maintenance cost;
- equipment (fixed equipment and specific construction) maintenance cost;
- insect and rodent control cost;
- incinerator and trash removal cost;
- building internal cleaning cost;
- special mechanical systems operations and maintenance cost;
- elevator, escalator, and dumbwaiter operations cost;
- plumbing and sewage systems operations and maintenance cost;
- fire protection systems maintenance cost; and
- air conditioning and ventilating system operations and maintenance cost.

4.4.3 Specific Life Cycle Cost Model III

This model is concerned with estimating the life cycle cost of an early warning radar system. The radar life cycle cost is expressed by [6]

$$LCC_r = C_p + C_o + C_s \tag{4.19}$$

where
LCC_r is early warning radar life cycle cost.
C_p is radar procurement cost.
C_o is radar operation cost.
C_s is radar logistic support cost.

The radar procurement cost, C_p, is expressed by

$$C_p = FC + ICC + DC + DOC \tag{4.20}$$

where
  FC is fabrication cost.
  ICC is installation and checkout cost.
  DC is design cost.
  DOC is document cost.

The radar operation cost, C_o, is defined by

$$C_o = C_1 + C_2 + C_3 \tag{4.21}$$

where
  C_1 is fuel cost.
  C_2 is cost of personnel.
  C_3 is cost of power.

The radar logistic support cost, C_s, is expressed by

$$C_s = CRL + CRM + CIS + CRS + CIT + AC \tag{4.22}$$

where
  CRL is cost of repair labor.
  CRM is cost of repair material.
  CIS is cost of initial spares.
  CRS is cost of replacement spares.
  CIT is cost of initial training.
  AC is age cost.

The life cycle cost predicted breakdown percentages for a specific early warning radar are available in Dhillon [6].

4.4.4 Specific Life Cycle Cost Model IV

This model is concerned with estimating the life cycle cost of inertial systems. The inertial systems life cycle cost is expressed by [20]

$$LCC_{is} = RDTC + PC + OMC \tag{4.23}$$

where
  LCC_{is} is inertial systems life cycle cost.
  $RDTC$ is research, development, test, and evaluation cost.
  PC is procurement cost.
  OMC is operation and maintenance cost.

The research, development, test, and evaluation cost, *RDTC*, is composed of eight elements:

- software cost;
- testing cost;
- program management cost;
- cost of conceptual studies;
- cost of engineering change proposals;
- cost of design engineering;
- cost of technical data; and
- training cost.

The 12 distinct components of the procurement cost include:

- cost of new facilities;
- cost of spares;
- support equipment acquisition cost;
- system recurring acquisition cost;
- cost of technical data;
- initial training course cost;
- training equipment cost;
- cost of production tooling and test equipment;
- production program start-up cost;
- cost of initial item management;
- field engineering cost; and
- equipment installation cost.

The operation and maintenance cost, *OMC*, is expressed by

$$OMC = \sum_{j=1}^{3} \sum_{i=1}^{n} OMC_{ji} \tag{4.24}$$

where
   n is total number of years.
   OMC_{ji} is operation and maintenance cost at the jth level of maintenance in the ith year.

Additional information on the model is available in DeBurkarte [20].

4.4.5 Specific Life Cycle Cost Model V

This model is concerned with estimating the life cycle cost of software. Sometimes, the model is called the "Boeing C-14 model" [21,22]. The software life cycle cost is expressed by [21,22]

$$LCC_s = AC_s + SC_s \tag{4.25}$$

where
  LCC_s is life cycle cost of software.
  AC_s is acquisition cost of software.
  SC_s is support cost of software.

The support cost of software, SC_s, is expressed by

$$SC_s = \left[(2.5)(LC) \sum SMM_j \right] (1+\alpha) + SC_a \tag{4.26}$$

where
  LC is direct labor cost per man-month.
  ΣSMM_j is required man-months for support in month j.
  α is overhead factor.
  SC_a is additional (other) support costs.

Additional information on the model is available in references 21 and 22.

4.5 Cost Estimation Methods

Over the years, many methods have been developed to estimate costs [23–26]. Some of the methods considered useful for application in the area of life cycle costing are presented next.

4.5.1 Cost Estimation Method I

This method is considered quite useful to obtain quick approximate cost estimates for similar new plants, projects, or equipment of different capacities. The cost-capacity relationship is defined by

$$C_n = C_o \left[\frac{K_n}{K_o} \right]^{\alpha} \tag{4.27}$$

where

  C_n is cost of the new plant, project, or equipment under consideration.
  C_o is cost of the old but similar equipment, plant, or project.
  K_n is capacity of the new plant, project, or equipment.
  K_o is capacity of old but similar equipment, plant, or project.
  α is the cost-capacity factor whose frequently used value is 0.6. The pro-
       posed values for this factor for items such as heat exchangers, heat-
       ers, pumps, and tanks are 0.6, 0.8, 0.6, and 0.7, respectively [23,27,28].

Example 4.1

An electric utility spent $900 million to construct a 1,000 megawatt (MW) nuclear power generating station. In order to satisfy the increasing demand for electricity, the company is planning to construct a 2,000 MW nuclear power generating station. Calculate the cost of the new station, if the value of the cost-capacity factor is 0.6.

By substituting the given data values into Equation (4.27), we get

$$C_n = 900 \left[\frac{2,000}{1,000} \right]^{0.6}$$

$$= \$1,364.15 \text{ million}$$

Thus, the construction cost of the new nuclear power station will be $1,364.15 million.

4.5.2 Cost Estimation Method II

This method is known as the Lang factor method, after its originator, H. L. Lang [29]. The method is used for obtaining quick order-of-magnitude cost estimates by utilizing historical average cost factors. Lang proposed to estimate total plant costs from the delivered equipment cost by using three factors as multipliers: $n = 3.10$ (for solid process plants), $n = 3.63$ (for solid-fluid plants), and $n = 4.74$ (for fluid process plants) [29].

Thus, the total estimate for plant cost is obtained by using

$$TPC = (n)(DEC) \tag{4.28}$$

where

  TPC is the total estimate for plant cost.
  n is the Lang factor, whose value depends on the nature of the plant.
  DEC is delivered equipment cost.

Example 4.2

Assume that a fluid-processing plant's delivered equipment cost is $40 million. Calculate the total plant cost.

By substituting the given data value and information into Equation (4.28), we get

$$TPC = (4.74)(40)$$

$$= \$189.6 \text{ million}$$

Thus, the total plant cost will be $189.6 million.

4.5.3 Cost Estimation Method III

This method is basically a refinement of the Lang factor method and is known as the Hand method, after its originator, W. E. Hand [30]. In the refinement, Hand proposed the use of different factors for various groups of equipment.

The total installed cost for each equipment group is defined by [25,30]

$$IC_t = (m)(DEC) \tag{4.29}$$

where
  IC_t is total installed cost of each equipment group.
  m is the Hand factor that covers field materials (structures, insulation, piping, electrical, finishes, and foundations), labor, and indirect costs. The values of the Hand factor for various groups of equipment are 2 (fired heaters), 2.5 (compressors), 2.5 (miscellaneous equipment), 3.5 (heat exchangers), 4 (pumps), 4 (pressure vessels), 4 (instruments), and 4 (fractionating towers).
  DEC is delivered equipment cost.

Note that the Hand factors do not incorporate a contingency allowance. Additional information on this method is available in references 25, 30, and 31.

4.5.4 Cost Estimation Method IV

This method is quite useful to make an order-of-magnitude approximation of operating labor requirements in the absence of a Manning table. The method is known as the Wessell method. Thus, the Wessell equation is expressed by [32]

$$\frac{OH}{\lambda} = \alpha \left[\frac{K}{(P)^{0.76}} \right] \tag{4.30}$$

where
  OH is number of operating man-hours.
  λ is tons of product.
  K is total number of process steps.
  P is capacity expressed in tons per day.

The values of α are 23 (for a batch operation with maximum labor), 10 (for a well-instrumented continuous process operation), and 17 (for an operation with average labor requirements). Additional information on this method is available in Humphreys [32].

4.5.5 Cost Estimation Method V

This method is known as the turnover ratio method and is considered the most efficient approach to estimating plant costs. However, it is probably the least accurate. The turnover ratio is defined by [6,32]

$$TOR = \frac{AS}{I} \tag{4.31}$$

where
    TOR is turnover ratio.
    AS is gross annual sales.
    I is fixed capital investment.

The gross annual sales, AS, is expressed by

$$AS = (SP)(PR) \tag{4.32}$$

where
    SP is unit selling price.
    PR is yearly production rate.

Note that the value of the turnover ratio, TOR, usually varies from around 0.2 to 8.

Example 4.3

Assume that a factory is to manufacture 50,000 units/year of a certain product. The selling price of a unit is $500. Calculate the fixed capital investment, if the turnover ratio is 4.

By substituting Equation (4.32) into Equation (4.31) and then substituting the given data values into the resulting equation, we get

$$4 = \frac{(500)(50,000)}{I} \tag{4.33}$$

By rearranging Equation (4.33), we obtain

$$I = \frac{(500)(50,000)}{4}$$

$$= \$6.25 \text{ million}$$

Thus, the fixed capital investment for the factory is $6.25 million.

Problems

1. Write an essay on life cycle cost models and cost estimation methods.
2. Discuss three types of life cycle cost models.
3. Write down life cycle cost equations for two general life cycle cost models.
4. Write down life cycle cost equations for two specific life cycle cost models.
5. Compare the general life cycle cost models with the specific life cycle cost models.
6. Write down a life cycle cost equation for switching power supplies.
7. What is the "Boeing C-14 model"?
8. Discuss the following two types of cost estimation methods:
   - the Hand method
   - the Wessell method
9. A solid-processing plant's delivered equipment cost is $20 million. Calculate the total plant cost by using the Lang factor method.
10. An electric power generation company spent $1,500 million to construct a 600 MW nuclear power generating station. In order to meet the increasing demand for electricity, the company is planning to construct a 1,500 MW nuclear power generating station. Calculate the cost of the new station, if the value of the cost-capacity factor is 0.7.

References

1. Dhillon, B. S. 1980. Life cycle cost: A survey. *Microelectronics and Reliability: An International Journal* 20:737–742.
2. Dhillon, B. S. 1983. *Reliability engineering in system design and operation.* New York: Van Nostrand Reinhold Company.
3. Gupta, Y. P. 1983. Life cycle cost models and associated uncertainties. In *Electronic systems effectiveness and life cycle costing,* ed. J. K. Skwirzyski, 535–549. Berlin: Springer–Verlag.
4. Dover, L. E., and Oswald, B. E. 1974. A summary and analysis of selected life cycle costing techniques and models. Master's thesis, Air Force Institute of Technology, Wright-Patterson Air Force Base, Ohio.
5. Sherif, Y. S., and Kolarik, W. J. 1981. Life cycle costing concepts and practice. *OMEGA* 9:287–296.
6. Dhillon, B. S. 1989. *Life cycle costing: Techniques, models, and applications.* New York: Gordon and Breach Science Publishers.
7. Goldman, A. S., and Slattery, T. B. 1967. *Maintainability: A major element of system effectiveness.* New York: John Wiley & Sons.
8. Kolarik, W. J. 1977. Analysis theory and procedures for determining and predicting availability, availability cost, and intangible effects for farm machinery systems. PhD dissertation, Oklahoma State University, Stillwater.

9. Siewiorek, D. P., and Swarz, R. S. 1982. *The theory and practice of reliable system design, digital press.* Bedford, MA: Digital Equipment Corporation.

10. Reiche, H. 1980. Life cycle cost. In *Reliability and maintainability of electronic systems,* ed. J. E. Arsenault and J. A. Roberts, 3–23. Potomac, MD: Computer Science Press, Potomac.

11. Locks, M. O. 1978. Maintainability and life cycle costing. *Proceedings of the Annual Reliability and Maintainability Symposium* 251–253.

12. Naval Weapons Engineering Support Activity, Naval Material Command. 1977. Life cycle cost guide for major weapon systems. Department of Defense, Washington, D.C.

13. Earles, M. 1981. *Factors, formulas, and structures for life cycle costing.* Concord, MA: Eddins–Earles.

14. Stordahl, N. C., and Short, J. L. 1968. The impact and structure of life cycle costing. *Proceedings of the Annual Symposium on Reliability* 509–515.

15. Blanchard, B. S. 1978. *Design and manage to life cycle cost.* Portland, OR: M/A Press.

16. Department of the Army. 1976. Research and development cost guide for Army material systems. Pamphlet no. 11-12. Department of Defense, Washington, D.C.

17. Department of the Army. 1976. Investment cost guide for Army material systems. Pamphlet no. 11-13. Department of Defense, Washington, D.C.

18. Department of the Army. 1976. Operating and support cost guide for Army material systems. Pamphlet no. 11-14. Department of the Army, Washington, D.C.

19. Monteith, D., and Shaw, B. 1979. Improved R, M, and LCC for switching power supplies. *Proceedings of the Annual Reliability and Maintainability Symposium* 262–265.

20. DeBurkarte, D. E. 1976. A standard life cycle cost model for inertial systems. *Proceedings of the National Aerospace and Electronics Conference* 687–695.

21. Ferens, D. V., and Harris, R. L. 1979. Avionics computer software operation and support cost estimation. *Proceedings of the IEEE National Aerospace and Electronics Conference* 296–300.

22. Boeing Aerospace Company. 1978. Advanced avionics systems for multimission applications. Report (vol. II)—Appendix G, Seattle, WA.

23. Desai, M. B. 1981. Preliminary cost estimating for process plants. *Chemical Engineering* July: 65–70.

24. Hackney, J. W. 1970. Estimating methods for process industry capital costs. In *Modern cost-engineering techniques,* ed. H. Popper, 43–58. New York: McGraw–Hill Book Company.

25. Ward, T. J. 1986. Cost-estimating methods. *Modular instruction series G: Design of equipment (plant design and cost estimating),* vol. 1, ed. J. Beckman, 12–21. New York: American Institute of Chemical Engineers.

26. Ostwald, P. F. 1974. *Cost estimating for engineering and management.* Englewood Cliffs, NJ: Prentice Hall, Inc.

27. Jelen, F. C., and Black, J. H., eds. 1983. *Cost and optimization engineering.* New York: McGraw–Hill Book Company.

28. Dieter, G. E. 1983. *Engineering design.* New York: McGraw—Hill Book Company.

29. Lang, H. J. 1947. Simplified approach to preliminary cost estimates. *Chemical Engineering* 54:130–133.
30. Hand, W. E. 1958. From flow sheet to cost estimate. *Petroleum Refiner* 37:331–334.
31. Wroth, W. F. 1960. Factors in cost estimation. *Chemical Engineering* 67:204–206.
32. Humphreys, K. K., ed. 1984. *Project and cost engineers' handbook*, 51–74. New York: Marcel Dekker, Inc.

5

Reliability, Quality, Safety, and Manufacturing Costing

5.1 Introduction

Reliability, quality, safety, and manufacturing costs play an important role in the total cost of engineering products. Therefore, they must be considered with care. Reliability cost is an important factor in any reliability program associated with an engineering product. It is associated with activities such as reliability allocation, prediction, and testing [1].

Quality costs usually form a significant component of the selling price of an engineering product. They cross department lines by involving various company activities such as design, manufacturing, purchasing, and service. Safety costs are becoming an important element of the economy. For example, in 1995, the cost of workplace accidents in the United States was estimated to be around $75 billion [2]. Needless to say, safety costs are associated with areas such as lawsuits, insurance, analysis, and corrective measures.

The manufacturing cost may be described as the sum of fixed and variable costs chargeable to the manufacture of a given product or item. Usually, this cost (i.e., manufacturing cost) excludes the costs associated with corporate administration, selling, research and development, and transportation and distribution.

This chapter presents various important aspects of reliability, quality, safety, and manufacturing costing.

5.2 Reliability Cost Classifications

Reliability cost may be categorized under the following four classifications [3]:

- *Prevention cost* includes items such as hourly and overhead rates for design engineers, reliability engineers, material engineers, technicians, and test and evaluation personnel; hourly cost and overhead rates for reliability screens; cost of yearly reliability training per capita; and cost of preventive maintenance programs.

- *Appraisal cost* involves items such as cost for vendor audit, new vendor qualification, and new part qualification; hourly and overhead rates for reliability evaluation, reliability demonstration, reliability qualification, environmental testing, and life testing; cost of test result reports; and average cost per part of assembly testing, auditing, screening, inspection, and calibration.
- *Internal failure cost* is composed of items such as cost of replaced parts or components; cost of spare part inventory; hourly and overhead rates for failure analysis, retesting, and troubleshooting and repair; and cost of production change administration.
- *External failure cost* includes items such as cost of liability assurance, cost of warranty administration and reporting, cost of failure analysis, cost of spare part inventory, cost of service kit, cost of replaced parts, and cost to repair a failure.

5.3 Models for Estimating Costs of Reliability-Related Tasks

Over the years, many mathematical models have been developed to estimate man-hours required to perform reliability-related tasks and, in turn, the cost of performing such tasks. Some of these models are presented next [1,4,5].

5.3.1 Model I

This model is concerned with estimating the total number of man-hours required to perform reliability prediction. This number is expressed by [1,4,5]

$$TMH_p = (4.54)\alpha^2\,\theta^2\,\beta \tag{5.1}$$

where
   TMH_p is total number of man-hours required to perform reliability prediction.
   α is the factor whose value depends on the type of report required: $\alpha = 1$ means an internal report is required; $\alpha = 2$ means a formal report is required.
   θ is the integer factor whose value varies from 1 to 3 depending on the level of detail: 1 = prediction exists, 2 = prediction is to be performed using similar system data, and 3 = full MIL-HDBK-217 [6] stress prediction is needed.
   β is the integer factor whose values vary from 1 to 4 depending on the percentage of commercial hardware used in the system or item under consideration: 1 = 76–100%, 2 = 51–75%, 3 = 26–50%, and 4 = 0–25%.

5.3.2 Model II

This model is concerned with estimating the total number of man-hours required to perform the reliability testing task. The number is defined by [1,4,5]

$$TMH_t = (182.07)(HCF) \qquad (5.2)$$

where

TMH_t is the total number of man-hours required to perform the reliability testing task.

HCF is the integer factor whose value varies from 1 to 3 depending on the degree of the hardware complexity: 1 = parts or components that are less than 15,000; 2 = parts or components that are 15,000–25,000; and 3 = parts or components that are greater than 25,000.

5.3.3 Model III

This model is concerned with estimating the total number of man-hours required for preparing the reliability and maintainability program plan. This number is defined by [1,4,5]

$$TMH_{pp} = (2.073)\gamma^2 \qquad (5.3)$$

where

TMH_{pp} is total number of man-hours required to prepare the reliability and maintainability program plan.

γ is number of MIL-STD-785/470 [6] tasks required. The recommended minimum and maximum values of γ are 4 and 22, respectively.

5.3.4 Model IV

This model is concerned with estimating the number of man-hours required for performing failure modes and effect analysis (FMEA). The number of man-hours required to perform this task is defined by [1,4,5]

$$TMH_f = (17.79)n \qquad (5.4)$$

where

TMH_f is total number of man-hours needed to perform FMEA.

n is total number of unique items requiring FMEA (e.g., number of circuit cards for piece-part and circuit-level FMEA or the number of pieces of equipment for equipment-level FMEA). The recommended minimum and maximum values of n are 3 and 206, respectively.

5.3.5 Model V

This model is concerned with estimating the total number of man-hours required to perform reliability allocation and modeling. The number of man-hours needed to carry out this task is defined by [1,4,5]

$$TMH_{am} = (4.05)(CF_{am})K \qquad (5.5)$$

where

TMH_{am} is total number of man-hours required to perform reliability allocation and modeling.

CF_{am} is the allocation and modeling complexity. The recommended values of the CF_{am} are 1, 2, and 3 for a series system, simple redundancy, and very complex redundancy, respectively.

K is total number of items in the allocation process. The recommended minimum and maximum values of K are 7 and 445, respectively.

5.4 Quality Cost Classifications and Their Distribution in the Industrial Sector

Quality costs may be divided into four classifications, as shown in Figure 5.1 [7]: prevention cost, appraisal cost, internal failure cost, and external failure cost. Each of these classifications is described separately in the following sections.

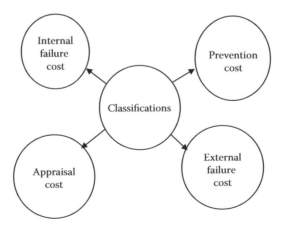

FIGURE 5.1
Quality cost classifications.

5.4.1 Prevention Cost

This cost is basically concerned with planning, implementing, and maintaining the quality system and is expressed by

$$C_p = C_{qe} + C_{qt} + C_{qp} + C_{qd} \qquad (5.6)$$

where
  C_p is prevention cost.
  C_{qe} is cost of quality engineering. This is concerned with the development and implementation of the inspection plan, the overall quality plan, etc.
  C_{qt} is cost of quality training. This includes the cost of developing and maintaining quality-related training programs.
  C_{qp} is cost of quality planning by functions excluding quality control.
  C_{qd} is cost associated with design and development of quality control and measurement equipment.

5.4.2 Appraisal Cost

This cost is concerned with determining the degree of conformance to quality-related specifications. It has many elements, including the cost of conducting product quality audits, the cost of testing and inspection of incoming material, the cost of materials and services consumed in testing, the cost of inspection and testing of items being manufactured, and the cost of maintenance and calibration of equipment used for evaluating quality.

5.4.3 Internal Failure Cost

This cost occurs when manufactured items fail to meet specified quality requirements prior to their ownership transfer to customers. The subcategories of the internal failure cost include repair cost, failure analysis cost, scrap cost, and re-inspection and retest cost.

5.4.4 External Failure Cost

This cost occurs when manufactured items fail to meet quality specifications after their delivery to customers. The external failure cost is expressed by

$$C_{ef} = C_r + C_w + C_a + C_h \qquad (5.7)$$

where
  C_{ef} is external failure cost.
  C_r is cost of repairing returned items.
  C_w is cost of warranties.
  C_a is cost of adjusting complaints.
  C_h is cost of replacement and handling of rejected (returned) items.

Although the distribution of quality costs may vary from one industrial sector to another and from one organization to another, their distribution in the banking and the electronic equipment manufacturing industries is as follows [8,9]:

- *Banking industry:* In this area, quality costs account for approximately 25% of a bank's total operating costs. The estimates of their distribution among prevention, appraisal, internal failure, and external failure cost classifications are 2, 28, 41, and 29% (of the total quality cost), respectively.

- *Electronic equipment manufacturing industry:* In this area, quality costs account for around 14% of the sales. The estimates of their distribution among prevention, appraisal, internal failure, and external failure cost classifications are approximately 45, 36, 13, and 6% (of the total quality cost), respectively.

5.5 Quality Cost Indexes and Quality Cost Reduction Approach

Many organizations use various types of quality cost indexes to monitor their performance. The values of such indexes are plotted on a periodic basis and their trends are monitored. Three of these indexes are presented next [10–13].

Index I is defined by

$$\theta_1 = \left[\frac{(C_q)(100)}{V_o} \right] + 100 \tag{5.8}$$

where
θ_1 is quality cost index.
C_q is total quality cost.
V_o is value of output.

The values of this index (θ_1) may be interpreted as follows [14]:

- $\theta_1 = 105$ can readily be achieved in a real-life environment.
- $\theta_1 = 110$–130 occurs in companies where the quality costs are totally ignored.
- $\theta_1 = 100$ means that there is absolutely no defective output.

Index II is defined by

$$\theta_2 = \frac{(C_q)(100)}{T_s} \tag{5.9}$$

where T_s is the total sales. Note that θ_2 is expressed as a percentage.

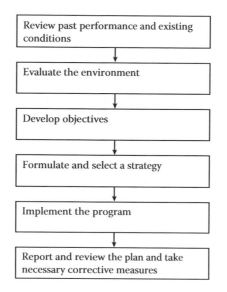

FIGURE 5.2
A six-step approach for reducing quality costs.

Index III is defined by

$$\theta_3 = \frac{(C_q)(100)}{C_d} \tag{5.10}$$

where C_d is the direct labor cost. Note that θ_3 is also expressed as a percentage. Usually, this index is used to eliminate inflation effects.

Although quality costs can be reduced in many different ways, the six-step approach shown in Figure 5.2 is considered quite useful for this purpose [14]. Additional information on the approach is available in Williams [14].

5.6 Safety Cost and Its Related Facts and Figures

Nowadays, the cost of safety has become an important factor in the life cycle cost of many engineering systems. Each year, the cost of safety in general is increasing at a significant rate. Some of the safety cost–related facts and figures are as follows:

- In 2000, work-related injuries cost the United States around $131 billion [15].
- The cost of the accident in 1979 at the Three Mile Island nuclear power plant was estimated to be approximately $4 billion [16].

- In 1993, a Virginia jury awarded $8 million to a worker for a back injury caused by a piece of equipment that fell [16].
- In 1996, a Paris-bound Trans World Airlines jet crashed due to a fuel-tank fire and killed all persons on board. A subsequent task force concluded that adding nonflammable gases (fuel-tank inverting) would decrease the risk of fuel-tank explosions quite significantly, but recommended against such changes because of the cost of between $10 billion and $20 billion [17].
- In 1997, three workers sued a computer equipment manufacturer for musculoskeletal disorders (MSDs) because they firmly believed that these disorders were due to keyboard entry activities [16]. The workers were awarded around $5.8 million.

5.7 Safety Cost Estimation Models

Over the years, many models to estimate safety cost have been developed. Some of these models are presented next.

5.7.1 Model I

This model is concerned with estimating the safety cost of a product over its life span and is expressed by [16,18]

$$LCSC_p = C_1 + C_2 + C_3 + C_4 - R \qquad (5.11)$$

where
$LCSC_p$ is product life cycle safety cost.
C_1 is cost of an accident prevention program.
C_2 is cost of insurance.
C_3 is recall cost.
C_4 is program cost.
R is reimbursements.

5.7.2 Model II

This is another mathematical model that can also be used to estimate the safety cost of a product over its life span. This is expressed by [16,18]

$$LCSC_p = SC_1 + SC_2 + SC_3 + SC_4 \qquad (5.12)$$

where
$LCSC_p$ is product life cycle safety cost.
SC_1 is safety cost during the product research and development phase. This cost is associated with the safety-related studies performed during this phase.

SC_2 is safety cost during the product production and construction phase. This cost is associated with the safety-related measures taken during this phase.

SC_3 is safety cost during the product operation and support phase. This cost is associated with safety-related activities performed during this phase.

SC_4 is safety cost during the product retirement and disposal phase. This cost is associated with safety-related actions taken to dispose of the product.

5.7.3 Model III

This model is concerned with estimating the total hidden cost of an accident and is expressed by [18]

$$AHC = C_d + C_m + C_{hsp} + C_{uw} + C_{iw} + C_e + C_{sp} + C_{nw} + C_{um} + C_{ro} \qquad (5.13)$$

where

AHC is total hidden cost of an accident.

C_d is cost of damage to equipment or material.

C_m is miscellaneous cost.

C_{hsp} is cost of time spent by clerical and higher supervisory personnel.

C_{uw} is cost of wages paid to uninjured workers for the time lost.

C_{iw} is cost of wages paid to injured workers for the time lost.

C_e is extra cost of overtime work necessitated by the accident under consideration.

C_{sp} is cost of wages paid to supervisory individuals for their time spent on activities necessitated by the accident under consideration.

C_{nw} is cost of the learning period required by new workers replacing injured workers.

C_{um} is uninsured medical cost borne by the organization or company.

C_{ro} is wage cost due to reduction in output of injured individuals after their return to work.

5.7.4 Model IV

This model is concerned with estimating total safety cost, which is defined by [16,18]

$$C_{st} = ILC + PMC + WIC + IC + IMC + MIC + LIC + RRC \qquad (5.14)$$

where

C_{st} is total safety cost.

ILC is cost of immediate losses due to accidents.

PMC is cost of accident prevention measures.

WIC is cost of welfare-related issues.

IC is cost of insurance.

IMC is cost associated with the immeasurable.

MIC is cost of miscellaneous safety-related issues.
LIC is cost of safety-related legal issues.
RRC is rehabilitation and restoration cost.

5.8 Manufacturing Costs

Manufacturing costs form a significant proportion of the life cycle cost of engineering products, equipment, and systems. They may be broken down into five categories, as shown in Figure 5.3 [19,20]. For new processes, the elements of the direct manufacturing cost include [19,20]:

- maintenance and repair cost;
- labor cost;
- cost of utilities;
- packaging and shipping cost;
- raw materials cost;
- royalties (if applicable);
- direct overhead cost (usually plant supervision);
- laboratory charges (process control and quality control work);
- factory supplies (house supplies, wiping cloths, instrument charts, etc.); and
- development cost (if applicable).

Similarly, the elements of the indirect manufacturing cost are plant indirect overhead cost (e.g., plant office expense), property taxes, depreciation, and insurance [19,20].

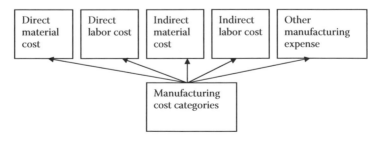

FIGURE 5.3
Manufacturing cost categories.

5.9 Manufacturing Cost Estimation Models

Over the years, many mathematical models have been developed to estimate various types of manufacturing cost. Some of these models are presented next.

5.9.1 Model I

This model is concerned with estimating the direct cost of material used in manufacturing, which is expressed by [20,21]

$$C_{dm} = (W)(P)\left[1 + \sum_{j=1}^{3} \alpha_j\right] - P_s \tag{5.15}$$

where

C_{dm} is direct material cost of a unit.
W is weight of a unit, usually expressed in pounds.
P is price of material expressed per linear foot, per pound, or per volume.
α_j is jth losses expressed in decimals for $j = 1$ (due to shrinkage), $j = 2$ (due to scrap), and $j = 3$ (due to waste).
P_s is unit price of expected material salvage expressed in dollars per unit.

Additional information on this model is available in Ostwald [21].

5.9.2 Model II

This model is concerned with estimating machining cost, which is expressed by [22–24]

$$MC = \frac{1}{60}\left[\frac{C_m(1+r_1)}{100} + \frac{r_2(1+r_3)}{100}\right](MT + T_i) \tag{5.16}$$

where

MC is machining cost.
C_m is machine cost expressed in dollars per hour.
r_1 is machine overhead rate expressed in percentage.
r_2 is operator labor rate expressed in dollars per hour.
r_3 is overhead rate of the operator expressed in percentage.
MT is machining time.
T_i is nonproduction or idle time.

Additional information on this model is available in references 22–24.

5.9.3 Model III

This model is concerned with estimating the tool cost associated with a cutting tool brazed to the tool holder. This cost is defined by [22]

$$C_t = \frac{(RSC)\beta + TC}{(\beta + 1)} \tag{5.17}$$

where
  C_t is tooling cost associated with a cutting tool brazed to the tool holder.
  RSC is cost associated with resharpening expressed in dollars.
  β is number of resharpenings.
  TC is tool cost expressed in dollars.

In the case of a throwaway (insert) tool, the cost, C_t, is expressed by

$$C_t = \frac{THC}{n_1} + \frac{TIC}{n_2} \tag{5.18}$$

where
  THC is cost of the tool holder.
  n_1 is total number of cutting edges in the life of the tool holder.
  TIC is tool insert cost expressed in dollars.
  n_2 is number of cutting edges.

5.9.4 Model IV

This model is concerned with estimating the average unit cost for a single-point, rough-turning operation. This cost is expressed by [21]

$$C_a = HC + TC + MC + THC \tag{5.19}$$

where
  C_a is average unit cost for a single-point, rough-turning operation.
  HC is handling cost.
  TC is tool cost.
  MC is machining cost.
  THC is tool-changing cost.

Equations for estimating HC, TC, MC, and THC follow.
  The handling cost, HC, is expressed by

$$HC = (T_h)(OTC) \tag{5.20}$$

where
  T_h is total handling time per work piece expressed in minutes.
  OTC is total operating time cost expressed in dollars per minute.

The tool cost, TC, is expressed by

$$TC = \frac{(WPT_m)(C_e)}{MTL} \tag{5.21}$$

where
WPT_m is machining time of work piece expressed in minutes per piece.
C_e is tool cost expressed as dollars per cutting edge.
MTL is mean tool life expressed in minutes.

The machining cost, MC, is expressed by

$$MC = (OTC)(WPT_m) \tag{5.22}$$

Finally, the tool-changing cost, THC, is expressed by

$$THC = \frac{(OTC)(WPT_m)(T_c)}{MTL} \tag{5.23}$$

where T_c is the tool-changing time expressed as minutes per operation. Additional information on this model is available in Ostwald [21].

Problems

1. List and discuss reliability cost classifications.
2. Write an essay on reliability, quality, and safety costing.
3. What are the quality cost classifications?
4. Compare quality cost classifications with reliability cost classifications.
5. Discuss an approach that can be used to reduce quality costs.
6. List at least five safety cost-related facts and figures.
7. List and discuss manufacturing cost categories.
8. Define at least two quality cost indexes.
9. Define a mathematical model that can be used to estimate the total hidden cost of an accident.
10. Compare reliability cost with safety cost.

References

1. RADC reliability engineer's toolkit. 1988. Published by the Systems Reliability and Engineering Division, Rome Air Development Center (RADC), Air Force Systems Command (AFSC), Griffiss Air Force Base, Rome, NY.

2. Spellman, F. R., and Whiting, N. E. 1999. *Safety engineering: Principles and practice.* Rockville, MD: Government Institutes.

3. Grant Ireson, W., and Coombs, C. F., eds. 1988. *Handbook of reliability engineering and management.* New York: McGraw–Hill Book Company.

4. National Technical Information Service (NTIS). 1987. R and M program cost drivers. Report no. RADC-TR-87-50 (ADA 182773), Springfield, VA.

5. Dhillon, B. S. 2005. *Reliability, quality, and safety for engineers.* Boca Raton, FL: CRC Press.

6. Department of Defense. 1998. Reliability program for systems and equipment. MIL-STD-785, Washington, D.C.

7. American Society for Quality Control. 1980. *Guide for managing vendor quality costs.* Milwaukee, WI.

8. Harrington, H. J. 1987. *Poor-quality cost.* New York: Marcel Dekker, Inc.

9. Breeze, J. D. 1980. Quality costs can be sold. *Proceedings of the American Society for Quality Control Conference* 795–801.

10. Evans, J. R., and Lindsay, W. A. 1989. *The management and control of quality.* New York: West Publishing Company.

11. Lester, R. H., Enrick, N. L., and Mottely, H. E. 1977. Quality control for profit. New York: Industrial Press.

12. Carter, C. L. 1978. *The control and assurance of quality, reliability, and safety.* Richardson, TX: C. L. Carter and Associates.

13. Sullivan, E., and Owens, D. A. 1983. Catching a glimpse of quality costs today. *Quality Progress* 16 (12): 21–24.

14. Williams, R. J. 1982. Guide for reducing quality costs. *Proceedings of the American Society for Quality Control Annual Conference* 360–366.

15. National Safety Council (NSC). 2001. Report on injuries in America in 2000. Itasca, IL: Author.

16. Hammer, W., and Price, D. 2001. *Occupational safety management and engineering.* Upper Saddle River, NJ: Prentice Hall, Inc.

17. Williams, W. E. 2001. Safety at all costs (www.worldnetdaily.com). Cave Junction, OR, September 5: 1–3.

18. Dhillon, B. S. 2003. *Engineering safety: Fundamentals, techniques, and applications.* River Edge, NJ: World Scientific Publishing.

19. *Chemical Engineering* (CE) cost file 92. 1964. Estimating manufacturing costs for new processes. *Chemical Engineering* August 17: 160–162.

20. Dhillon, B. S. 1989. *Life cycle costing: Techniques, models, and applications.* New York: Gordon and Breach Science Publishers.

21. Ostwald, P. F. 1974. *Cost estimating for engineering and management.* Englewood Cliffs, NJ: Prentice Hall Inc.

22. Dieter, G. E. 1983. *Engineering design: A materials and processing approach.* New York: McGraw–Hill Book Company.

23. Boothroyd, G. 1975. *Fundamentals of metal machining and machine tools.* New York: McGraw–Hill Book Company.

24. Armarego, E. J. A., and Brown, R. H. 1969. *The machining of metals.* Englewood Cliffs, NJ: Prentice Hall, Inc.

6

Maintenance, Maintainability, Usability, and Warranty Costing

6.1 Introduction

Each year billions of dollars are spent to produce various types of engineering products. Past experiences indicate that, in many cases, the cost of procuring an engineering product is less than the cost of ownership over its life span. According to Blanchard, Verma, and Peterson [1], the hidden costs related to equipment operation and support can account for as high as 75% of the equipment life cycle cost. Maintenance, maintainability, usability, and warranty costs play an important role in the life cycle cost of an engineering product. Therefore, careful consideration must be given to estimating such costs.

The maintenance cost may be described simply as the labor and materials expense required for maintaining engineering products in suitable use condition. In some systems—particularly military systems—the maintenance cost can be as high as 70% of life cycle costs [2]. Maintainability is an important factor in the total cost of equipment because increase in maintainability can result in reduction in equipment operation and support costs. Thus, maintainability costs are basically concerned with equipment design.

Usability costs are concerned with a wide range of activities employed in developing effectively usable engineering products. Some examples of these activities are establishing a definition for end user requirements, developing specifications for usability objectives, performing task analysis, and conducting usability testing. Warranty costs occur when engineering equipment manufacturers provide buyers with written statements guaranteeing the integrity of their equipment. The responsibilities of the manufacturers are outlined by these statements in situations when their equipment happens to be defective.

This chapter presents various important aspects of maintenance, maintainability, usability, and warranty costing.

6.2 Reasons for Maintenance Costing, Factors Influencing Maintenance Cost, and Types of Maintenance Costs

There are many reasons for maintenance costing. Some of the important ones include [3]:

- to prepare budgets;
- to make equipment replacement decisions;
- to control costs;
- to compare maintenance costs' effectiveness with industry averages;
- to identify maintenance cost drivers;
- to improve productivity;
- to compare competing maintenance methods;
- to provide appropriate inputs in the design of new equipment or items; and
- to perform equipment or item life cycle cost studies.

Some of the important factors influencing maintenance costs are shown in Figure 6.1 [3,4].

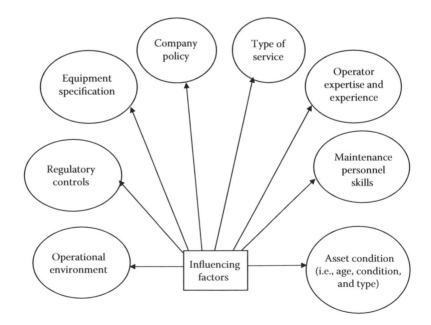

FIGURE 6.1
Factors influencing maintenance costs.

There are basically two main categories of maintenance costs: preventive maintenance cost and corrective maintenance cost. The former is concerned, directly or indirectly, with actions performed on a planned, periodic, and specific schedule for keeping a piece of equipment or item in stated working condition through the process of rechecking and reconditioning. More specifically, these actions are precautionary measures undertaken to forestall or decrease the probability of failures or an unacceptable level of degradation in subsequent service, rather than rectifying failures after their occurrence.

The corrective maintenance cost is directly or indirectly concerned with the unscheduled maintenance and repair to return equipment or items to a specified condition. These actions are carried out because involved maintenance personnel or users perceive deficiencies or failures.

6.3 Equipment Maintenance Cost

The maintenance cost of the entire ownership cycle of equipment is expressed by [4,5]

$$EMC_p = [\lambda_c (CMC) + \lambda_p (PMC)] \left[\frac{1 - (1 + j)^{-m}}{j} \right] \quad (6.1)$$

where

EMC_p is present value of the maintenance cost of the entire ownership cycle of equipment.
λ_c is constant corrective maintenance rate of equipment per year.
CMC is expected cost of a corrective maintenance action.
λ_p is constant preventive maintenance rate of equipment per year.
PMC is expected cost of a preventive maintenance action.
m is equipment expected life expressed in years.
j is annual interest rate.

Example 6.1

Assume that annual preventive and corrective maintenance rates of an engineering system are 5 and 2, respectively. Each preventive and corrective action costs $200 and $1,000, respectively. Calculate the present value of the system maintenance cost, if the expected system life and annual interest rate are 10 years and 5%, respectively.

By substituting the given data values into Equation (6.1), we get

$$EMC_p = [(2)(1,000) + (5)(200)] \left[\frac{1 - (1 + 0.05)^{-10}}{0.05} \right]$$

$$= \$23,165.20$$

Thus, the present value of the system maintenance cost is $23,165.20.

6.3.1 Maintenance Equipment Cost

This is expressed by [6]

$$MEC = C_{rd} + \alpha C_a \tag{6.2}$$

where
   MEC is maintenance equipment cost.
   C_{rd} is research and development cost associated with the maintenance equipment.
   α is number of pieces of maintenance equipment.
   C_a is maintenance equipment unit acquisition cost.

6.4 Preventive and Corrective Maintenance Labor Cost Estimation

The preventive maintenance labor cost is expressed by [7]

$$PMLC = (LR) \left[\frac{\sum_{j=1}^{K} f_j \, APMT_j}{\sum_{j=1}^{K} f_j} \right] \tag{6.3}$$

where
   $PMLC$ is equipment preventive maintenance labor cost.
   LR is hourly labor rate.
   K is number of data points.
   f_j is frequency of jth preventive maintenance action expressed in actions per operating hour, after adjustment for equipment duty cycle, for $j = 1, 2, 3,..., K$.
   $APMT_j$ is average time, in hours, required to carry out jth preventive maintenance action for $j = 1, 2, 3,..., K$.

Similarly, the corrective maintenance labor cost is given by [7]

$$CMLC = \frac{T_{so}(LC)(MTTR)}{MTBF} \tag{6.4}$$

where
   $CMLC$ is equipment annual corrective maintenance labor cost.
   T_{so} is equipment annual scheduled operating hours.
   LC is hourly corrective maintenance labor cost.
   $MTTR$ is equipment mean time to repair.
   $MTBF$ is equipment mean time between failures.

Example 6.2

Assume that a system is scheduled to operate for 2,500 hours annually and its mean time between failures and mean time to repair are 700 hours and 3 hours, respectively. Calculate the system annual corrective maintenance labor cost, if the hourly corrective maintenance labor cost is $30.

By inserting the given data values into Equation (6.4), we get

$$CMLC = \frac{(2500)(30)(3)}{700}$$

$$= \$321.43$$

Thus, the system annual corrective maintenance labor cost is $321.43.

6.5 Repair Manpower, Maintenance Material, and Spare and Repair Parts Costs

According to a U.S. military document [6], repair cost with respect to manpower can be estimated by using the following equation:

$$RMC = \theta(1 - F_{rs})RC_{um} \tag{6.5}$$

where

RMC is repair manpower cost.
θ is total number of repairable units failing over system or equipment life.
F_{rs} is repairable shrinkage factor due to damage, loss, etc. Its values are tabulated in reference 6 and vary from 0 to 0.1375.
RC_{um} is unit repair cost with respect to manpower.

The total number of repairable units failing over system or equipment life is expressed by

$$\theta = \lambda(T)(\alpha)(SL) \tag{6.6}$$

where

λ is item constant failure rate.
T is annual operating hours.
α is number of repairable items.
SL is system or equipment life (in reference 6, taken to be 10 years).

The unit repair cost with respect to manpower is given by

$$RC_{um} = (x)(F_{mu})(y) \tag{6.7}$$

where

x is mean number of man-hours per repair action.
F_{mu} is manpower use factor. Its values are tabulated in reference 6 and they vary from 1.04 to 3.
y is hourly manpower cost including overhead.

The maintenance material cost is an important element of the total maintenance cost. For example, according to Neibel [8], in the United States industrial sector, the cost of maintenance materials typically accounts for 40–50%
of the total maintenance cost. Because the cost of excessive inventory and
obsolete parts is an important factor in most maintenance stockrooms and
storerooms, well-planned and efficiently operated stockrooms and storerooms can help to reduce the cost of materials.

The total cost of stock or stores at the time of repair can be calculated by
using the following equation [8]:

$$SC_t = ITC + C_i + (W_{it} - ITC) + (0.01)(ITC)(t) + (0.1)(ITC)$$

$$= ITC + C_i + \left[\frac{(t)(ITC) + (10)(ITC)}{100} \right] \tag{6.8}$$

where
SC_t is total cost of stock or stores at the time of repair.
C_i is inventory cost per item.
ITC is present worth of the inventory item cost including procurement and
    delivery costs.
W_{it} is inventory item worth after K periods.
t is time, expressed in months, during which the stock item is in inventory.

Note that Equation (6.8) allows an inflation rate of 1% per month of procurement cost, while the item under consideration is in inventory, and 10% for
the item's total shelf life to take into consideration factors such as spoilage,
obsolescence, theft, and deterioration.

Equations to calculate W_{it}, C_i, and ITC, respectively, are presented next.

$$W_{it} = (ITC)(1 + j)^K \tag{6.9}$$

$$C_i = (S_b)(FSC)/(n)(y) \tag{6.10}$$

$$ITC = (1 + UL + SL)(PC)(WT) - MSP \tag{6.11}$$

where
j is interest rate for a given period.
K is total number of interest periods.
S_b is size of a bin expressed in square feet.
FSC is yearly floor space cost per square foot.
n is mean number of items stored in a bin.
y is reciprocal of total years that an item usually spends in inventory.
UL is amount of losses generated by unused stock returned to inventory
    considered too small in terms of quantity for use in the future.

SL is total amount of losses due to scrap, chips, skeletons, and so on.

PC is procurement cost or price (i.e., the delivered price) of material per unit.

WT is weight or other unit of quantity of material used.

MSP is unit price of material salvaged.

The spare and repair parts cost is defined by

$$SRC = ISRC + CC + DSRC + SSRC + OSRC \qquad (6.12)$$

where

SRC is spare and repair parts cost.

ISRC is intermediate spare and repair parts cost.

CC is total cost of consumables.

DSRC is depot spare and repair parts cost.

SSRC is supplier spare and repair parts cost.

OSRC is organizational spare and repair parts cost.

6.6 Maintenance Cost Estimation Models

Over the years, many mathematical models have been developed to estimate various types of maintenance-related costs. Four of these models are presented next.

6.6.1 Model I

This model is concerned with estimating equipment initial logistic support cost, which is expressed by [9]

$$EILSC = ISRC + ITHC + IMC + TDPC + LPMC + ITTEC + PC + OTSEPC \qquad (6.13)$$

where

EILSC is equipment initial logistic support cost.

ISRC is cost of initial spare and repair parts.

ITHC is initial transportation and handling cost.

IMC is cost of initial inventory management.

TDPC is technical data preparation cost.

LPMC is cost of logistic program management.

ITTEC is initial training and training equipment cost.

PC is provisioning cost, including preparation of procurement data for essential spares, test, and support equipment.

OTSEPC is procurement cost of operational test and support equipment.

6.6.2 Model II

This model is concerned with estimating software maintenance cost. This cost is expressed by [10]

$$SMC = \frac{3(m)(LC)}{\theta} \tag{6.14}$$

where
  SMC is software maintenance cost.
  m is total number of instructions to be changed per month.
  LC is labor cost per man-month.
  θ is difficulty constant. Its values for hard programs, easy programs, and
      programs of medium difficulty are 100, 500, and 250, respectively.

6.6.3 Model III

This model is concerned with estimating Doppler radar maintenance cost. This is expressed by [11]

$$DRMC = \frac{(C_a)y}{1000} \tag{6.15}$$

where
  $DRMC$ is Doppler radar maintenance cost.
  C_a is Doppler radar annual maintenance cost.
  y is number of years in service.

The natural logarithm of C_a is given by

$$ln\,C_a = \theta_1 + \theta_2\,ln\,C_{fu} \tag{6.16}$$

where
  $\theta_1 = -1.269$
  $\theta_2 = 0.696$
  C_{fu} is the first unit cost of the Doppler radar expressed in 1974 dollars ($\times 10^3$).

6.6.4 Model IV

This model is concerned with estimating the fire control radar maintenance cost, which is expressed by [11]

$$FCRMC = (MC_{ph})\,h_1\,h_2/1000 \tag{6.17}$$

where
  $FCRMC$ is fire control radar maintenance cost.
  MC_{ph} is maintenance cost per flying hour per unit expressed in 1974
      dollars ($\times 10^3$).
  h_1 is total number of annual flying hours.
  h_2 istotal number of years in service.

The natural logarithm of MC_{ph} is expressed by

$$ln\, MC_{ph} = \alpha_1 + \alpha_2\, ln\, P_{pw} \tag{6.18}$$

where
   $\alpha_1 = -2.086$
   $\alpha_2 = 0.611$
   P_{pw} is peak power expressed in kilowatts.

6.7 Maintenance Cost Data Collection

As various types of cost data are needed in maintenance costing, management decides the types of cost data the maintenance department should collect by considering their potential applications. Four types of maintenance cost-related data are collected [12]:

- *Labor costs* are usually obtained by using items such as timesheets, job tickets, and maintenance work orders.
- *Spare parts and supplies costs* are usually obtained from maintenance work orders.
- *Overhead costs* are usually obtained from the company accounting department.
- *Equipment costs* are usually obtained from either purchase orders or suppliers' invoices.

6.8 Maintainability Investment Cost Elements

The main elements of maintainability investment cost are as follows [6]:

- repair parts;
- system test and evaluation;
- new operational facilities;
- system engineering management;
- data;
- training;
- prime equipment; and
- support equipment.

Additional information on these elements is available in reference 6.

6.9 Manufacturer Warranty and Reliability Improvement Warranty Costs

The cost of warranty to an equipment manufacturer can be quite significant. It can be estimated by using the following equation [13]:

$$MWC = (C_{mu})(n)(\lambda) + C_{fw} \qquad (6.19)$$

where
  MWC is manufacturer or contractor warranty cost.
  C_{mu} is mean cost for the manufacturer or contractor to repair a unit sent back for warranty service.
  n is operating hours of equipment under warranty during the warranty period.
  λ is average constant failure rate per hour of equipment under warranty during the warranty period.
  C_{fw} is manufacturer or contractor warranty fixed cost.

The manufacturer warranty and reliability improvement warranty cost is expressed by [14]

$$MWRC = FC_m + C_{ia} + C_d + P + C_x \qquad (6.20)$$

where
  $MWRC$ is manufacturer warranty and reliability improvement warranty cost.
  FC_m is fixed cost of the manufacturer associated with the warranty.
  C_{ia} is cost associated with reliability improvement actions for attaining the achieved mean time between failures (i.e., reliability improvement warranty period average).
  C_d is cost of damages associated with not meeting the specified turnaround time.
  P is profit.
  $C_{x'}$ is cost.

The cost, $C_{x'}$ is expressed by

$$C_x = (C_{mr})(n)(T_w)(UR)/MTBF_a \qquad (6.21)$$

where
  C_{mr} is manufacturer's cost per unit repair.
  n is number of systems or items to be delivered.
  T_w is length of the warranty period.
  UR is usage rate expressed in operating time per calendar time.
  $MTBF_a$ is achieved mean time between failures (i.e., reliability improvement warranty period average).

6.10 Usability Costing and Related Facts and Figures

A wide range of activities is generally employed in effectively developing usable engineering products. The cost of these activities depends on factors such as the scope of the product under consideration, functional range, the number of scenarios to be studied, the number of users to be studied, and the skill and experience of the usability specialists [15]. Some of the usability costing-related facts and figures include:

- An American Airlines study reported that catching a usability-related problem early in the design process can decrease the cost of correcting it by 60–90% [16].
- A study revealed that the total training time for new users of a standard personal computer was approximately 21 hours as opposed to around 11 hours for users of a user-friendly computer [17].
- A study reported that the annual cost of lost productivity to American businesses is around $100 billion because office workers "futz" with their machines an average of 5.1 hours per week [18].
- A study revealed that approximately 80% of software maintenance cost is due to unmet or unforeseen user requirements [19].
- A study reported that an Australian insurance company spent approximately $100,000 (Australian) on a usability-related project concerned with redesigning its application forms to reduce customer errors and saved $536,023 (Australian) annually [20].

Additional information on usability-related facts and figures is available in Dhillon [21].

6.11 Principal Costs of Ignoring Product Usability and Product Usability Cost Estimation

The principal costs of ignoring engineering product usability are as follows [22]:

- *User error cost* is concerned with the users of engineering products making errors. In turn, these errors result in reduction in their productivity.
- *Poor productivity cost* consists of the additional time spent by engineering product users with products that are difficult to use.
- *Training cost* deals with the training of users when the product is first introduced. It increases significantly when products are difficult to use.

- *Customer support cost* involves a customer hotline telephone service, usually provided by product manufacturers for people having difficulties using the product. Past experiences indicate that products that are difficult to use generate greater customer or user requests for help. In turn, more people are required to handle users or customers, thus resulting in greater customer support cost.
- *Poor sale cost* involves dissatisfied customers or users not purchasing the product in the future, even when they are made aware of improvements in product usability. Past experiences indicate that a dissatisfied customer or user influences roughly 10 others to avoid buying the product in question [22].
- *Tarnished corporate image cost* is concerned with users or customers buying not only the current or improved usability version of the product in question, but also other products manufactured by the same firm.

The product usability cost can be estimated by using the following equation when usability cost data are available for similar products of different capacities [23]:

$$DPC_u = SPC_u \left[\frac{CP_d}{CP_o} \right]^\theta \tag{6.22}$$

where
DPC_u is desired product usability engineering cost.
SPC_u is known usability engineering cost of a similar item, product, or piece of equipment of known capacity CP_o.
CP_d is desired product capacity.
θ is cost-capacity factor. The value of this factor varies for different products or items. In circumstances when no data for θ are available, it is considered quite reasonable to assume its value to be 0.6.

Example 6.3
Assume that the usability engineering cost of an 80 GB computer system is $300. Calculate the cost of usability engineering of a similar 100 GB computer system if the value of the cost-capacity factor is 0.8.
    By substituting the specified data values into Equation (6.23), we get

$$DPC_u = (300) \left[\frac{100}{80} \right]^{0.8}$$

$$= \$358.63$$

Thus, the cost of usability engineering of the similar 100 GB computer system is $358.63.

Problems

1. What are the principal reasons for maintenance costing?
2. Discuss at least seven factors that influence maintenance cost.
3. Assume that annual preventive and corrective rates of an engineering system are 7 and 3, respectively. Each preventive and corrective action costs $400 and $1,200, respectively. Calculate the present value of the system maintenance cost, if the expected system life and annual interest rate are 12 years and 3%, respectively.
4. Discuss the collection of three types of maintenance cost-related data.
5. What are the principal elements of maintainability investment cost?
6. Discuss equipment warranty cost.
7. What is the equipment usability cost?
8. List at least five usability costing-related facts and figures.
9. Discuss the costs of ignoring engineering product usability.
10. Assume that an engineering system is scheduled to operate for 3,000 hours annually and its mean time between failures and mean time to repair are 800 hours and 4 hours, respectively. Calculate the system annual corrective maintenance labor cost, if the hourly corrective maintenance labor cost is $40.

References

1. Blanchard, B. S., Verma, D., and Peterson, E. L. 1995. *Maintainability: A key to effective serviceability and maintenance management.* New York: John Wiley & Sons.
2. Dhillon, B. S. 1989. *Life cycle costing: Techniques, models, and applications.* New York: Gordon and Breach Science Publishers.
3. Levitt, J. 1997. *The handbook of maintenance management.* New York: Industrial Press.
4. Dhillon, B. S. 2002. *Engineering maintenance: A modern approach.* Boca Raton, FL: CRC Press.
5. Dhillon, B. S. 1996. *Engineering design: A modern approach.* Chicago, IL: Richard D. Irwin.
6. Department of the Army. 1976. Engineering design handbook: Maintainability engineering theory and practice, AMCP 706-133, Department of Defense, Washington, D.C.
7. Department of the Army. 1975. Engineering design handbook: Maintenance engineering techniques. AMCP 706-132. Department of Defense, Washington, D.C.
8. Neibel, W. B. 1994. *Engineering maintenance management.* New York: Marcel Dekker.

9. Dhillon, B. S. 1999. *Engineering maintainability.* Houston, TX: Gulf Publishing Company.
10. Sheldon, M. R. 1979. *Life cycle costing: A better method of government procurement.* Boulder, CO: Westview Press.
11. Cost analysis of avionics equipment, vol. 1. 1974. Prepared by U.S. Air Force Systems Command, Wright-Patterson Air Force Base, Ohio. The NTIS report no. AD741132. Available from the National Technical Information Service (NTIS), Springfield, VA.
12. Jordan, J. K. 1990. *Maintenance management.* Denver, CO: American Water Works Association.
13. Balaban, H. S., and Meth, M. A. 1978. Contractor risk associated with reliability improvement warranty. *Proceedings of the Annual Reliability and Maintainability Symposium* 123–129.
14. Gates, R. K., Bicknell, R. S., and Bortz, J. E. 1977. Quantitative models used in the RIW decision process. *Proceedings of the Annual Reliability and Maintainability Symposium* 229–236.
15. Rosson, M. B., and Carroll, J. M. 2002. *Usability engineering: Scenario-based development of human-computer interaction.* San Francisco: Morgan Kaufmann Publishers.
16. Laplante, A. 1992. Put to the test. *Computerworld* 27 (July 27): 75–77.
17. Nielson, J. 1993. *Usability engineering.* Boston: Academic Press, Inc.
18. Westlake Consulting Company. 1997. SBT Accounting Systems. Houston, TX.
19. Pressman, R. S. 1992. *Software engineering: A practitioner's approach.* New York: McGraw–Hill Book Company.
20. Fisher, P., and Sless, D. 1990. Information design methods and productivity in the insurance industry. *Information Design Journal* 6 (2): 103–129.
21. Dhillon, B. S. 2004. *Engineering usability: Fundamentals, applications, human factors, and human error.* Stevenson Ranch, CA: American Scientific Publishers.
22. Keinonen, T., Mattelmaki, T., Soosalu, M., and Sade, S. 1997. Usability design methods. Technical report, Department of Product and Strategic Design, University of Art and Design, Helsinki, Finland.
23. Dieter, G. E. 1983. *Engineering design: A materials and processing approach,* 324–366. New York: McGraw–Hill Book Company.

7

Computer System Life Cycle Costing

7.1 Introduction

Today computers play an important role in our daily lives. Over the years, their applications have increased quite dramatically, ranging from personal use to controling nuclear reactors and space systems. The computer industry has become an important component of the global economy; a vast sum of money is spent to produce, operate, and maintain computers each year. For example, in fiscal year 1980, the U.S. government spent over $57 billion on computer systems [1].

Computer systems are made up of both hardware and software components and the percentage of overall computer system cost spent on hardware has changed quite remarkably over the years. For example, in 1955, the hardware component accounted for 80% of total computer system cost; however, in 1985, the hardware component cost decreased to just 10% [2]. This means that, nowadays, software cost is a very important element of total computer system cost—more specifically, the computer system life cycle cost.

Over the years, many models and procedures have been developed to estimate directly or indirectly computer system life cycle cost. This chapter presents various important aspects of computer system life cycle costing.

7.2 Computer System Life Cycle Cost Models

A number of mathematical models are used to estimate life cycle cost of a computer system. Two such models are presented next [3,4].

Model I divides the life cycle cost of a computer system into two main components: procurement cost and ownership cost. Thus, the life cycle cost of a computer system is expressed by

$$LCC_{CS1} = C_p + C_0 \tag{7.1}$$

where
LCC_{CS1} is computer system life cycle cost.
C_p is computer system procurement cost.
C_0 is computer system ownership cost.

The procurement cost includes the cost of items such as system hardware, software license fees, installation, training, and documentation. Similarly, the ownership cost includes the cost of items such as preventive maintenance, computer downtime, supplies, and corrective maintenance.

Model II is a more detailed model to estimate the life cycle cost of a computer system. However, the model assumes that the cost of corrective maintenance is the only ownership cost of the computer system. Thus, the life cycle cost of the computer system is defined by

$$LCC_{CS2} = C_{p2} + \sum_{j=1}^{n} (C_{mj}) \alpha_j / (1+i)^j \tag{7.2}$$

where
LCC_{CS2} is life cycle cost of the computer system.
C_{p2} is procurement cost of the computer system.
n is computer system expected life expressed in years.
C_{mj} is corrective maintenance cost of a single maintenance activity during year j.
α_j is expected number of times that the computer system will fail during year j.
i is discount rate.

For the same number of computer system failures occurring in each year, Equation (7.2) simplifies to

$$LCC_{CS2} = C_{p2} + \alpha \sum_{j=1}^{n} C_{mj} / (1+i)^j \tag{7.3}$$

where α is expected number of computer system failures per year.

Example 7.1
Assume that the procurement cost and expected useful life of a computer system are $4,000 and 5 years, respectively. The computer system's expected number of failures per million hours is 100 and its only ownership cost is the cost of corrective maintenance. Calculate the life cycle cost of the computer system, if the cost of each corrective maintenance call is $200 and the yearly discount or interest rate is 4%.

The expected number of failures of the computer system per year is given by

$$n = \frac{(100)(8,760)}{1,000,000} = 0.876 \ failures/year$$

Using the preceding calculated value and the given data in Equation (7.3), we get

$$LCC_{CS2} = 4,000 + (0.876)(200) \sum_{j=1}^{5} (1+0.04)^{-j}$$

$$= \$4,779.96$$

Thus, the life cycle cost of the computer system is \$4,779.96.

7.3 Computer System Maintenance Cost

The computer system maintenance cost is an important component of computer system life cycle cost. This section presents two mathematical models to estimate, directly or indirectly, computer system maintenance cost.

Model I is concerned with estimating the maintenance cost of computer system hardware, which is expressed by [3,4]

$$C_{sm} = C_{pm} + C_{cm} + C_i \tag{7.4}$$

where
C_{sm} is monthly maintenance cost of the computer system hardware.
C_{pm} is preventive maintenance cost of the computer system hardware.
C_{cm} is corrective maintenance cost of the computer system hardware.
C_i is inventory cost.

The preventive maintenance cost of the computer system hardware is expressed by

$$C_{pm} = (OH)\theta[SPMT_e + TT_e]/SPM_i \tag{7.5}$$

where
OH is equipment operating hours per month.
θ is hourly rate of the customer engineer. This also includes the spare parts usage rate.
$SPMT_e$ is customer engineer's scheduled preventive maintenance time.
TT_e is customer engineer's travel time to perform preventive maintenance.
SPM_i is scheduled preventive maintenance interval.

The customer engineer's hourly rate, θ, is expressed by

$$\theta = \frac{PR(1+OR)}{\alpha} + C_p \tag{7.6}$$

where

PR is hourly pay rate of the customer engineer.

OR is overhead rate.

C_p is cost of parts per hour.

α is fraction of time that the customer engineer spends on the maintenance activity. Note that the customer engineer spends the remaining fraction of time on items such as paperwork, training, and waiting.

The corrective maintenance cost of the computer system hardware is defined by

$$C_{cm} = \theta(OH)[TT_{ecm} + MTTR]/MTBF \tag{7.7}$$

where

TT_{ecm} is customer engineer's travel time for performing corrective maintenance.

$MTTR$ is mean time to repair.

$MTBF$ is mean time between failures.

The inventory cost, C_i, is expressed by

$$C_i = (V_{ip})(R_i) \tag{7.8}$$

where

V_{ip} is value of the maintenance spare parts inventory.

R_i is monthly inventory cost rate, which includes items such as handling cost, interest charges for spares, and depreciation.

Model II is concerned with estimating the annual labor cost of servicing a computer system. This cost depends on many factors, including average cost of labor, mean time to preventive maintenance, preventive maintenance time interval, mean time between failures, and mean time to repair.

The annual labor cost is defined by

$$C_a = (LC_h)(8760)\left[\frac{(T_{apm} + TT_{pm})}{T_{bpm}} + \frac{(TT_r + MTTR)}{MTBF}\right] \tag{7.9}$$

where

LC_h is labor cost per hour.

T_{apm} is average time taken to perform preventive maintenance.

TT_{pm} is travel time associated with a preventive maintenance call.

T_{bpm} is mean time between preventive maintenance services.

TT_r is travel time associated with a repair or corrective maintenance call.

$MTTR$ is mean time to repair.

$MTBF$ is mean time between failures.

Example 7.2

Assume the following data values concerning servicing a computer system:

$LC_h = \$50$
$T_{apm} = 4$ hours
$TT_{pm} = 0.5$ hour
$T_{bpm} = 2,500$ hours
$TT_r = 1$ hour
$MTTR = 2$ hours
$MTBF = 3,000$ hours

Calculate the annual labor cost for servicing the computer system by using Equation (7.9).

By substituting the given data values into Equation (7.9), we get

$$C_a = (50)(8760)\left[\frac{(4+0.5)}{2,500} + \frac{(1+2)}{3,000}\right]$$

$$= \$1,226.4$$

Thus, the annual labor cost for servicing the computer system is \$1,226.40.

7.4 Software Costing and Related Difficulties

Over the years, software cost has increased to a very high level from a rather low percentage of the total computer system cost. For example, according to a U.S. Air Force study conducted in 1972, cost of software in 1955 accounted for less than 20% of total computer system cost; however, its projection for 1985 was around 80% of the total amount [3]. Furthermore, in July 1976, *Newsweek* magazine reported that the ratios of computer system hardware cost to software cost were 1:4 and 4:1 in 1976 and the 1950s, respectively.

Needless to say, today software cost has become a very important element of the computer system life cycle cost. Over time, many methods and procedures have been developed to estimate software cost.

Some of the difficulties faced in estimating software cost include [5]:

- poor understanding of the effects of management and technical-related constraints;
- poor understanding of the software development and maintenance processes;
- unavailability of adequate historic data to make appropriate checks;

- unavailability of adequate historic data for calibration applications (A calibration may be described as a process through which a model is fitted to a given cost estimating condition.); and

- project-to-project comparison inhibition because of firm belief in a project's uniqueness.

7.5 Software Life Cycle Cost Influencing Factors and Model

Many factors influence software life cycle cost. They may be grouped under five distinct attributes [6]:

- *Group I: computer attributes.* Some examples of these attributes are turnaround time, speed, and storage constraints.

- *Group II: project attributes.* Some examples of these attributes are schedule constraints, use of software tools, and modern programming practices [7].

- *Group III: size attributes.* Some examples of these attributes are numbers of inputs, outputs, data elements, and instructions.

- *Group IV: product attributes.* Some examples of these attributes are required software reliability, the choice of programming language, and software product complexity.

- *Group V: personnel attributes.* These attributes affect software cost much more than any other groups of attributes. Some examples of personnel attributes are teamwork; experience with respect to items such as programming language, applications, and virtual machines; and personnel and team capabilities.

The life cycle cost of software is composed of seven distinct elements, as shown in Figure 7.1. This is expressed mathematically by [8]

$$LCC_{se} = SDC + SAC + CC + SOSC + SIC + STIC + SDOC \tag{7.10}$$

where
LCC_{se} is software life cycle cost.
SDC is software design cost.
SAC is software analysis cost.
CC is software code and checkout cost.
$SOSC$ is software operating and support cost.
SIC is software installation cost.
$STIC$ is software test and integration cost.
$SDOC$ is software documentation cost.

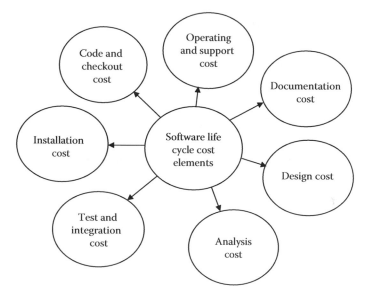

FIGURE 7.1
Software life cycle cost elements.

Table 7.1 presents the main elements of costs of software design, analysis, operating and support, code and checkout, test and integration, installation, and documentation [8].

7.6 Software Cost Estimation Methods and Models

Over the years, many methods and models have been developed to estimate software cost. Some of these methods and models are presented next.

7.6.1 Software Cost Estimation Methods

Many methods have been used to estimate software costs, including [3,9]:

- algorithmic models;
- top-down estimating;
- bottom-up estimating;
- analogy; and
- expert opinion.

The algorithmic models are described later in detail, and additional information on the remaining four methods is available in Dhillon [3] and Boehm [9].

The algorithmic models may be described as the models that provide at least one mathematical algorithm to generate a computer software cost estimate as a

TABLE 7.1

Subelements of Software Life Cycle Cost Elements

No.	Software Life Cycle Cost Element	Subelements
1	Design cost	Cost of flow charts
		Data structure cost
		Cost of test procedures
		Cost of input and output parameters
2	Analysis cost	Cost of system requirements
		Cost of program requirements
		Cost of design requirements and specifications
		Cost of interface requirements
3	Operating and support cost	Cost of modifications
		Cost of test revisions
		Cost of documentation revisions
		Cost of environments
4	Code and checkout cost	Cost of desk checks
		Cost of coded instructions
		Cost of compiling programs
5	Test and integration cost	Cost of program test
		Cost of system integration
6	Installation cost	Validation cost
		Verification cost
		Certification cost
7	Documentation cost	Cost of listings
		Cost of user manual
		Cost of maintenance manual

function of several variables. These variables are considered very important cost drivers. The five common types of algorithmic models are presented next [9].

7.6.1.1 Tabular Models

Generally, these models are quite straightforward to comprehend and implement. They are composed of tables relating cost driver variables' values to portions of the software development effort or to multipliers employed to adjust the effort estimate. Three examples are the Black et al. [10], Aron [11], and Wolverton [12] models.

7.6.1.2 Composite Models

These models incorporate an amalgamation of four types of functions (i.e., linear, tabular, analytic, and multiplicative) for determining software effort as a function of cost driver variables. Past experiences indicate that composite models are relatively more difficult to learn and use, in addition to requiring more data and effort.

7.6.1.3 Analytic Models

These models take the following form [3]:

$$ET = f(x_1, x_2, ..., x_n) \tag{7.11}$$

where
ET is effort.
f is a function (it is to be noted that this function is neither linear nor multiplicative).
x_j is cost driver variable j for $j = 1, 2, 3, ..., n$.
n is total number of cost driver variables.

Two good examples of the analytic models are the Putnam [13] and Halstead [14] models.

7.6.1.4 Linear Models

These models take the following form [3]:

$$ET = N_o + \sum_{j=1}^{K} N_j x_j \tag{7.12}$$

where
ET is effort.
K is total number of cost driver variables.
N_j is coefficient chosen to best fit the observed data points for $j = 0, 1, 2, 3, ..., K$.
x_j is cost driver variable j for $j = 1, 2, 3, ..., K$.

An important reference to the earliest use of the linear model is the System Development Corporation software cost-estimation study performed in the mid-1960s [15]. Finally, it was concluded that there are many nonlinear interactions in the software development process for linear models to perform very effectively.

7.6.1.5 Multiplicative Models

These models take the following form [3]:

$$ET = M_o \prod_{j=1}^{n} M_j^{x_j} \tag{7.13}$$

where
   ET is effort.
   n is total number of cost driver variables.
   M_j is coefficient chosen to best fit the observed data for $j = 0, 1, 2, 3,\ldots, n$.
   x_j is cost driver variable j for $j = 1, 2, 3,\ldots, n$.

Past experiences indicate that multiplicative models work fairly well for reasonable, independently selected variables. Additional information on multiplicative models is available in Walston [15] and Herd et al. [16].

7.6.2 Software Cost Estimation Models

As mentioned earlier, many types of models can be used to estimate software cost (see references 3, 5, 9, and 17). Some of these models used directly or indirectly to estimate software cost are presented next.
   Model I is concerned with estimating the software development cost, which is expressed by [16,18]

$$SDC = SPC + SSC \tag{7.14}$$

where
   SDC is software development cost.
   SPC is software primary development cost.
   SSC is software secondary development cost.

The software primary development cost is defined by

$$SPC = (ALR)(MR) \tag{7.15}$$

where
   ALR is average labor rate of the software development manpower expressed
         in dollars per man-month. It includes items such as administration
         cost, general cost, and overhead cost.
   MR is manpower required for software development expressed in man-
         months. This includes activities such as analysis, design, code, test,
         debug, and checkout.

Similarly, the software secondary development cost is expressed by

$$SSC = \sum_{i=1}^{m} C_i$$
$$= \lambda(SPC) \tag{7.16}$$
$$= \lambda(ALR)(MR)$$

where
  m is total number of secondary resources.
  C_i is cost associated with secondary resource i for $i = 1, 2, 3, ..., m$.
  λ is ratio of software secondary development cost to software primary development cost.

Model II is concerned with estimating the duration of a software project. The model predicts the minimum project duration under the assumption that the total hardware will be available during the project life. Thus, the minimum project duration is expressed by [19]

$$D_{min} = \frac{D_p}{S_a} \qquad (7.17)$$

where
  D_{min} is minimum project duration.
  D_p is total programmer-months.
  S_a is average staff size allocated to the software project under consideration.

Additional information on the model is available in Schneider [19].
  Model III is concerned with estimating the software marketing cost, which is expressed by [20]

$$C_{sm} = C_{ho} + C_{fs} \qquad (7.18)$$

where
  C_{sm} is annual software marketing cost.
  C_{ho} is cost associated with the home office.
  C_{fs} is field sales-related cost.

The field sales-related cost is given by

$$C_{fs} = [\theta(BS) + \theta(SAS)\alpha](1 + r_o) + r_c(APS) \qquad (7.19)$$

where
  θ is total number of people involved in sales.
  BS is annual base salary of a salesperson.
  SAS is annual salary of a system analyst.
  α is total number of system analysts employed per salesperson.
  r_o is overhead rate.
  r_c is commission rate.
  APS is annual product sales.

Model IV is concerned with estimating the software quality cost, which is expressed by [21]

$$SQC = PC + AC + IFC + EFC \qquad (7.20)$$

where

SQC is software quality cost.

PC is prevention cost associated with activities performed to prevent the occurrence of software errors. Some examples of these activities are developing a software quality infrastructure, improving and updating that infrastructure, and performing the regular activities necessary for its successful operation.

AC is appraisal cost associated with activities pertaining to the detection of software errors in software projects under consideration. Some examples of the appraisal cost components are the cost of software testing, cost of reviews, and cost of assuring quality of external participants (e.g., subcontractors).

IFC is internal failure cost associated with correcting software errors discovered through testing, design reviews, and acceptance tests prior to the installation of the software under consideration at customer sites.

EFC is external failure cost associated with correcting software failures discovered by customers after the installation of software at their sites.

Model V is concerned with estimating software project effort, in programmer-months, in situations when very little information about the project under consideration is available, except its expected delivery instructions. The software project effort is expressed by [19]

$$SPE = (1.7)(I_d)(S_{cf})(L_{af}) \tag{7.21}$$

where

SPE is software project effort expressed in programmer-months.

I_d is delivered instructions expressed in thousands.

S_{cf} is software complexity factor. Its values for trivial, moderately complex, and very complex software are 1, 5, and 10, respectively.

L_{af} is labor estimate adjustment factor expressed in decimal fraction. Its recommended values for rather poorly managed projects and under best conditions are 2.9 and 0.435, respectively.

Model VI was developed by the U.S. Naval Air Development Center and is concerned with estimating the effort to develop software [22]. This effort is expressed by [17,22]

$$SDE = 2.8x + 1.3y + 33z + 10K + L - 17M - 188 \tag{7.22}$$

where

SDE is total number of man-months needed for the software development.

x is delivered program's machine language instructions expressed in thousands.

y is contractor man-miles traveled.

z is total number of document types produced or generated.

K is total number of independent consoles in the delivered system.

L is number of new instructions in percentages.

M is average programmer experience with the system under consideration, expressed in years.

Problems

1. Write an essay on computer system life cycle costing.
2. Assume that the acquisition cost and expected useful life of a computer system are $5,000 and 6 years, respectively. The computer system's expected number of failures per million hours is 80 and its only ownership cost is the cost of corrective maintenance. Calculate the life cycle cost of the computer system, if the cost of each corrective maintenance call is $150 and the annual discount or interest rate is 6%.
3. Discuss the major difficulties faced in estimating software cost.
4. Discuss the factors that influence software life cycle cost.
5. What is the difference between computer hardware and software costing?
6. Write an equation that can be used to estimate software life cycle cost.
7. Discuss software cost estimation methods.
8. Compare tabular models with linear models with respect to software costing.
9. Discuss a mathematical model that can be used to estimate computer system hardware maintenance cost.
10. What is the main difference between Equation (7.2) and Equation (7.3)?

References

1. Carter faces problems in achieving his 1980 budget goals. 1979. *Wall Street Journal*, January 23: 4–5.
2. Keene, S. J. 1992. Software reliability concepts. *Annual Reliability and Maintainability Symposium Tutorial Notes* 1–21.
3. Dhillon, B. S. 1989. *Life cycle costing: Techniques, models, and applications.* New York: Gordon and Breach Science Publishers.
4. Phister, M. 1978. Analyzing computer technology costs—Part II: Maintenance. *Computer Design* October: 109–118.
5. Stanley, M. 1982. Software cost estimating, royal signals and radar establishment. Memorandum no. 3472. Procurement Executive, Ministry of Defense, Malvern, Worcs., UK.

6. Boehm, B. W. 1984. Software life cycle factors. In *Handbook of software engineering*, ed. C. R. Vick and C. V. Ramamoorthy, 494–518. New York: Van Nostrand Reinhold Company.

7. Dhillon, B. S. 1987. *Reliability in computer system design*. Norwood, NJ: Ablex Publishing Corporation.

8. Earles, M. E. 1981. *Factors, formulas, and structures for life cycle costing*. Concord, MA: Eddins–Earles.

9. Boehm, B. W. 1981. *Software engineering economics*. Englewood Cliffs, NJ: Prentice Hall, Inc.

10. Black, R. K. D., Curnow, R. P., Katz, R., and Gray, M. D. 1977. BCS software production data. Report no. RADC-TR-77-116. Boeing Computer Services, Inc. Available from the National Technical Information Services (NTIS), Springfield, VA.

11. Aron, J. D. 1969. Estimating resources for large programming systems, NATO Science Committee, Rome.

12. Wolverton, R. W. 1974. The cost of developing large-scale software. *IEEE Transactions on Computers* 23: 615–636.

13. Putnam, L. H. 1978. A general empirical solution to the macro software sizing and estimating problem. *IEEE Transactions on Software Engineering* 4: 345–361.

14. Halstead, M. H. 1977. *Elements of software science*. New York: Elsevier.

15. Walston, C. E., and Felix, C. P. 1977. A method of programming measurement and estimation. *IBM Systems Journal* 16:54–73.

16. Herd, J. R., Postak, J. N., Russell, W. E., and Stewart, K. R. 1977. Software cost estimation study—Study results. Report no. RADC-TR-77-220, vol. 1. Doty Associates, Inc., Rockville, MD.

17. James, T. G. 1977. Software cost estimating methodology. *Proceedings of the IEEE National Aerospace and Electronics Conference* 22–28.

18. Doty, D. L., Nelson, P. J., and Stewart, K. R. 1977. Software cost estimation study, vol. II. Report no. RADC-TR-77-220. Prepared by Doty Associates, Inc., Rockville, MD.

19. Schneider, V. 1978. Prediction of software effort and project duration—Four new formulas. *Sigplan Notices* 13:49–59.

20. Phister, M. 1976. *Data processing technology and economics*. Santa Monica, CA: Santa Monica Publishing Company.

21. Galin, D. 2004. *Software quality assurance*. Harlow, Essex, England: Pearson Education Ltd.

22. Buck, F. et al. 1971. A cost by function model for avionic computer systems. Report no. NADC-SD-7088, vol. 1. Developed by Naval Air Development Center, Warminster, PA.

8

Transportation System Life Cycle Costing

8.1 Introduction

Each year a vast amount of money is spent to develop, manufacture, and operate transportation systems such as motor vehicles, trains, aircraft, and ships throughout the world. This amount has become an important element of the global economy. Saving a small percentage of this amount can result in a large sum of money.

The concept of life cycle costing is increasingly being applied to make various types of decisions concerning transportation systems, particularly at their design and procurement stages. The main reason for the increasing use of the life cycle costing concept during a transportation systems' design and procurement stages is that past experiences indicate that many transportation systems' ownership costs (i.e., logistics and operating cost) often exceed their procurement costs. This is also the case for many other engineering products and systems. In fact, according to Ryan [1], the ownership costs of certain engineering products and systems can vary from 10 to 100 times their acquisition costs.

Over the years, a large number of publications have appeared on various aspects of transportation system life cycle costing. This chapter presents various important aspects of aircraft, ship, urban rail, and motor vehicle life cycle costing.

8.2 Aircraft Life Cycle Cost

Although the life cycle cost breakdown structure of an aircraft can vary from one organization to another and from one type of aircraft to another, it can be broken down into four parts as follows:

$$LCC_a = C_1 + C_2 + C_3 + C_4 \tag{8.1}$$

where

LCC_a is aircraft life cycle cost.
C_1 is aircraft research, development, test, and evaluation cost.
C_2 is aircraft production cost.
C_3 is aircraft initial support costs associated with items such as support equipment, spares, data, special equipment, and contractual training.

C_4 is aircraft operations and support cost associated with items such as base-level maintenance, training, and operations personnel and depot-level engine and component repair.

Usually, the life cycle cost of a typical fighter aircraft is broken into three categories [1]:

$$LCC_{fa} = DC_{fa} + PC_{fa} + OSC_{fa} \qquad (8.2)$$

where
LCC_{fa} is life cycle cost of a typical fighter aircraft.
DC_{fa} is fighter aircraft development cost.
PC_{fa} is fighter aircraft acquisition cost.
OSC_{fa} is fighter aircraft operations and support cost.

According to Huie and Harris [2], for a typical fighter aircraft, the operations and support cost, acquisition cost, and development cost over its life span of 15 years usually account for approximately 55, 35, and 10% of the life cycle cost, respectively.

The four main components of fighter aircraft development cost are design and development cost, test and evaluation cost, flight test support cost (e.g., cost of spares, ground support equipment, and personnel), and cost of data (e.g., test and stress reports). Normally, activities such as design, manufacturing, and testing account for roughly 90% of the development cost. The factors that drive the cost of the development include mission capabilities, physical characteristics such as weight, size, reliability, and maintainability characteristics (e.g., mean time between failures and mean time to repair).

The six main components of the fighter aircraft acquisition cost are shown in Figure 8.1 [2,3]. Two of these components (i.e., flyaway cost and cost of initial support) account for an extremely large percentage of the acquisition cost. The flyaway cost includes the cost of the airframe, engine, and avionics, and the cost of initial support includes the cost of spares, ground support equipment, inventory entry and management, and training and training equipment.

Some of the main drivers of the acquisition cost are reliability and maintainability characteristics, maintenance concept, mission capabilities, and training system requirements.

The fighter aircraft operations and support cost is composed of nine main components:

- cost of fuel;
- cost of personnel;
- cost of depot maintenance;
- cost of facilities;
- cost of base maintenance material;
- cost of modifications;

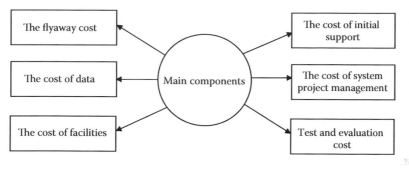

FIGURE 8.1
Main components of fighter aircraft acquisition cost.

- cost of replenishing spares;
- cost of replacement training; and
- cost of item management.

The five cost components that account for approximately 85% of the operations and support cost are fuel cost, depot maintenance cost, personnel cost, base maintenance material cost, and replenishing spares cost.

The seven factors that drive the fighter aircraft operations and support cost are shown in Figure 8.2 [3].

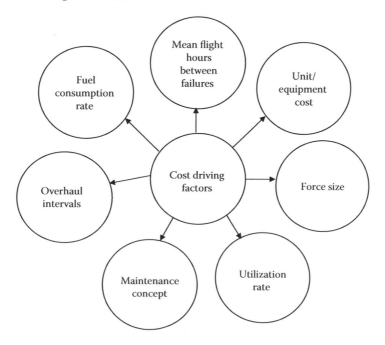

FIGURE 8.2
Fighter aircraft operation and support cost driving factors.

8.3 Aircraft Turbine Engine Life Cycle Cost

In the overall cost of an aircraft, the turbine engine is an important subsystem. The engine life cycle cost is expressed by [4,5]

$$LCC_{ae} = TEDC + TEPIC + TEPQC + TEBMC + TEDMC \qquad (8.3)$$

where
LCC_{ae} is aircraft turbine engine life cycle cost.
$TEDC$ is turbine engine development cost.
$TEPIC$ is turbine engine part improvement cost.
$TEPQC$ is turbine engine production quantity cost.
$TEBMC$ is turbine engine base maintenance cost.
$TEDMC$ is turbine engine depot maintenance cost.

Equations to estimate the five right-hand-side elements of Equation (8.3) are given in Jones [4] and Nelson [5].

8.4 Aircraft Cost Drivers

There are many aircraft cost drivers. In general, they may be grouped under the following three areas [2]:

- design;
- manufacturing; and
- operations and support.

The design cost drivers may be divided into three categories: reliability and maintainability requirements, performance requirements, and specifications. Two important elements of the reliability and maintainability requirements are mean time between failures (MTBF) and mean time to repair (MTTR).

Similarly, the four elements of the performance requirements are speed, payloads, range, and mission role. Finally, the elements of the specifications include corrosion control and fatigue life. Four main categories of the typical manufacturing cost drivers are shown in Figure 8.3 [2,3].

The elements of the material category include steel, aluminum, titanium, and composite. Some of the elements of the manufacturing process category are forgings, castings, machined parts, and sheet metal. The two main elements of the structure category are wing and body. The wing includes components such as the number of hard points, wet versus dry, and complexity of control surfaces. Similarly, the body includes components such as landing gear attachment and wing attachment.

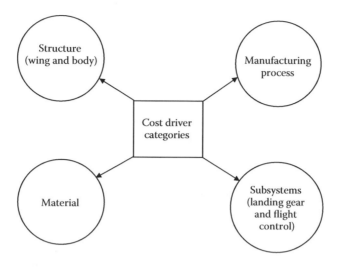

FIGURE 8.3
Main categories of typical aircraft manufacturing cost drivers.

The subsystems category has two main elements: flight control and landing gear. The flight control includes items such as the number of redundancies and mechanical versus fly-by-wire. Similarly, the landing gear includes items such as the number of wheels and brakes.

8.4.1 Helicopter Maintenance Cost Drivers

Many maintenance cost drivers are associated with helicopters. For example, maintenance cost drivers for military helicopters include the rotor system, power plants, transmissions, inspections, and others [3]. Generally, the breakdown percentages of direct maintenance cost (parts and labor) for these cost drivers are roughly 29, 27, 12, 9, and 23%, respectively. The breakdown percentages within the rotor system are blades (80%) and hub (20%).

Furthermore, note that the major contributor to the rotor hub operation and support cost is the seal leak, which results in lubricant loss and fluid. Similarly, two major contributors to the rotor blade operation and support cost are foreign object damage and inability to repair damaged blades.

8.4.2 Aircraft Airframe Maintenance Cost Drivers

According to a study performed in the early 1970s, the top nine airframe maintenance cost components were as follows [3]:

- brakes;
- tires;

- nose landing gear wheel and tire;
- flight control power control units;
- constant speed drive;
- motor-driven hydraulic pump;
- engine-driven hydraulic pump;
- auxiliary power unit; and
- starter.

8.4.3 Combat Aircraft Hydraulic and Fuel Systems Cost Drivers

The following are cost drivers of a combat aircraft hydraulic system [3,6]:

- valves (9%);
- pumps (26%);
- filters (12%);
- reservoirs (20%);
- accumulators (7%);
- plumbing (12%); and
- other (14%).

The cost drivers of a combat aircraft fuel system are valves (33%), pumps (27%), filters (8%), measurement (17%), and other (15%) [3,6].

8.5 Cargo Ship Life Cycle Cost

Ships are an important mode of transportation; over 90% of the world's cargo is transported by merchant ships. The life cycle cost of a cargo ship is expressed by [7]

$$LCC_{CS} = BC + OC + AC + OPC \tag{8.4}$$

where
   LCC_{CS} is cargo ship life cycle cost.
   BC is cargo ship building cost, including the cost of items such as machinery, outfitting, and hull.
   OC is the cargo ship owner's cost, including items such as naval architect's fee, attorney's fee, and consulting fees.
   AC is cargo ship accommodation cost, including the cost of items such as steel, hull engineering, and outfiting.

OPC is cargo ship operating cost, including the cost of items such as fuel, maintenance and repair, cargo handling, part changes, wages, inventory, protection and indemnity insurance, hull and machinery insurance, and subsistence.

Additional information on cargo ship life cycle cost is available in Earles [7].

8.6 Operating and Support Costs for Ships

Over the years, many formulas have been developed to estimate various types of ship operating and support costs. Some of these formulas that have been developed for the U.S. Navy are presented next [7,8].

8.6.1 Formula I

This formula is concerned with estimating the cost of repair parts:

$$C_{rp} = A + (B)(LD) \tag{8.5}$$

where
C_{rp} is cost of repair parts expressed per steaming hour (i.e., underway and not underway) in 1976 dollars.
$A = 28.083$
$B = 0.00263$
LD is full load displacement expressed in tons.

8.6.2 Formula II

This formula is concerned with estimating the cost of conventional fuel and is expressed by

$$C_{cf} = D + (E)(HP) - (F)x \tag{8.6}$$

where
C_{cf} is cost of conventional fuel.
$D = 166.021$
$E = 0.001974$
HP is total shaft horsepower.
$F = 490.220$
x is a dummy variable whose value is either 1 (when the ship is nuclear powered) or 0 (when the ship is not nuclear powered).

8.6.3 Formula III

This formula is concerned with estimating the ship overhaul cost, which is expressed by

$$C_{Sho} = (MD_r)(N) + (0.25)(MD_r)(N) \tag{8.7}$$

where
  C_{Sho} is ship overhaul cost.
  MD_r is repair man-days per overhaul.
  $N = \$150$ (in 1976 dollars)

Note that the right-hand side of Equation (8.7) is composed of two components: labor cost $[(MD_r)(N)]$ and material cost $[(0.25)(MD_r)(N)]$.

8.6.4 Formula IV

This formula is concerned with estimating ship supplies' cost. This is expressed by

$$C_{SS} = G + (H)(\alpha) + (I)x \tag{8.8}$$

where
  C_{SS} is ship supplies' cost.
  $G = 44{,}797.515$
  $H = 248.260$
  α is ship crew size (i.e., officers + enlisted individuals).
  $I = 478{,}830$
  x is a dummy variable whose value is either 1 (i.e., when the ship is nuclear powered) or 0 (i.e., when the ship is not nuclear powered).

Note that Equation (8.8) provides the annual cost of health, safety, and welfare supplies expressed in 1976 dollars.

8.7 Urban Rail Life Cycle Cost

Urban rail is an important means of transportation around the globe. Each day it transports millions of passengers and millions of dollars worth of goods from one place to another. The urban rail life cycle cost is defined by [7]

$$LCC_{ur} = SCC + SOC \tag{8.9}$$

where
  LCC_{ur} is life cycle cost.
  SCC is capital cost, including the cost of items such as vehicles, track and track work, power substations and distribution, stations, and yard and maintenance facilities.
  SOC is operating cost, including the cost of items such as power, transportation-associated manpower, and maintenance of tracks, vehicles, and equipment.

Additional information on this topic is available in references 3, 7, and 9.

8.8 Car Life Cycle Cost

Each day, a vast sum of money is spent to procure and operate various types of cars throughout the world. The life cycle cost of a car is defined by [3,10]

$$LCC_C = C_a + \sum_{j=1}^{n} OC_j + SMC_j + USMC_j + C_d \qquad (8.10)$$

where

LCC_C is life cycle cost of the car.

C_a is acquisition cost.

n is expected life of the car expressed in years.

OC_j is operating cost (i.e., for gas, oil, tires, etc.) for year j for $j = 1, 2, 3,..., n$.

SMC_j is scheduled maintenance cost (i.e., for tune-up, lubrication, etc.) for year j for $j = 1, 2, 3,..., n$.

$USMC_j$ is unscheduled maintenance or repair cost (dependent on car failure rate) for year j for $j = 1, 2, 3,..., n$.

C_d is car disposal plus any other cost.

Additional information on car life cycle costing is available in references 3, 7, and 10.

Example 8.1

Assume that the acquisition cost of a car is $23,000. Annual scheduled and unscheduled maintenance costs are $200 and $400, respectively. Furthermore, the annual operating cost of the car is $1,500 and its life expectancy is 7 years. Calculate the car life cycle cost, if its disposal cost and the annual interest rate are $2,000 and 4%, respectively.

By using an equation given in Chapter 2 and in reference 3 and the given data, we get the following present values of the car operating cost, scheduled maintenance cost, and unscheduled maintenance cost, respectively:

$$OC_p = 1500 \left[\frac{1 - (1 + 0.04)^{-7}}{0.04} \right] = \$9,003.1$$

$$SMC_p = 200 \left[\frac{1 - (1 + 0.04)^{-7}}{0.04} \right] = \$1,200.4$$

and

$$USMC_p = 400 \left[\frac{1 - (1 + 0.04)^{-7}}{0.04} \right] = \$2400.8$$

where

OC_p is present value of operating cost.

SMC_p is present value of scheduled maintenance cost.

$USMC_p$ is present value of unscheduled maintenance cost.

Using an equation given in Chapter 2 and in reference 3 and the specified data values, we get the following present value of the car disposal cost:

$$C_{dp} = \frac{2,000}{(1+0.04)^7} = \$1,519.8$$

where C_{dp} is present value of the car disposal cost.

Using all of the preceding calculated values, the given data value, and Equation (8.10), we get the following value for the car life cycle cost:

$$LCC_C = \$23,000 + \$9,003.1 + \$1,200.4 + 2,400.8 + \$1,519.8$$

$$= \$37,124.1$$

Thus, the car life cycle cost is $37,124.10.

8.9 City Bus Life Cycle Cost Estimation Model

A bus is an important means of transport in cities throughout the world. This model is concerned with estimating city bus cost over the life span of the vehicle. Thus, the city bus life cycle cost is expressed by [3,11]

$$LCC_{Cb} = VAC + TC + IOC + WC + LC + FC$$
$$+ MCC + RC + GOC + OHC + CIC + TC \tag{8.11}$$

where
 LCC_{Cb} is city bus life cycle cost.
 VAC is vehicle acquisition cost.
 TC is tire cost.
 IOC is cost of intermediate overhauls.
 WC is cost of wages.
 LC is lubricant cost.
 FC is fuel cost.
 MCC is cost of maintenance and checkup.
 RC is repair cost.
 GOC is cost of general overhauls.
 OHC is cost of overhead.
 CIC is cost of compulsory insurance.
 TC is taxes.

Additional information on the model is available in Zalud and Lanc [11].

Problems

1. Write an essay on transportation system life cycle costing.
2. What are the main components of fighter aircraft development cost?
3. What are the driving factors of fighter aircraft operation and support cost?
4. What are the main components of the fighter aircraft procurement cost?
5. Mathematically, define the life cycle cost of an aircraft turbine engine.
6. Discuss helicopter maintenance cost drivers.
7. What are the aircraft airframe maintenance cost drivers?
8. Mathematically, define the life cycle cost of a cargo ship.
9. Write formulas to estimate ship overhaul cost and conventional fuel cost.
10. Assume that the procurement cost of a car is $30,000. The annual scheduled and unscheduled maintenance costs are $300 and $500, respectively. Furthermore, the annual operating cost of the car is $1,000 and its life expectancy is 8 years. Calculate the car life cycle cost, if its disposal cost and the annual interest rate are $1,500 and 6%, respectively.

References

1. Ryan, W. J. 1968. Procurement views of life cycle costing. *Proceedings of the Annual Symposium on Reliability* 164–168.
2. Huie, E., and Harris, H. F. 1980. Balanced design—minimum cost solution. In *Design to cost and life cycle cost.* North Atlantic Treaty Organization (NATO) Advisory Group for Aerospace Research and Development (AGARD) conference proceedings no. 289, 13.1–13.14.
3. Dhillon, B. S. 1989. *Life cycle costing: Techniques, models, and applications.* New York: Gordon and Breach Science Publishers.
4. Jones, E. J. 1980. Design to life cycle cost interaction of engine and aircraft. In *Application of design to cost and life cycle cost to aircraft engines.* North Atlantic Treaty Organization (NATO) Advisory Group for Aerospace Research and Development (AGARD) lecture series no. 107, 3.1–3.15.
5. Nelson, J. R. 1980. An approach to the life cycle analysis of aircraft turbine engines. In *Application of design to cost and life cycle cost to aircraft engines.* North Atlantic Treaty Organization (NATO) Advisory Group for Aerospace Research and Development (AGARD) lecture series no. 107, 2.1–2.27.
6. Grieser, H. 1980. Impact on system design of cost analysis of specifications and requirements. In *Design to cost and life cycle cost.* North Atlantic Treaty Organization

(NATO) Advisory Group for Aerospace Research and Development (AGARD) conference proceedings no. 289, 6.1–6.10.

7. Earles, M. E. 1981. *Factors, formulas, and structures for life cycle costing.* Concord, MA: Eddins–Earles.

8. Eskew, H. L, Frazier, T. P., and Heilig, P. T. 1977. An operating and support cost model for aircraft carriers and surface combatants. Report no. ADA044744. Administrative Science Corporation, Alexandria, VA. Available from National Technical Information Service (NTIS), Springfield, VA.

9. Griffin, T. 2007. Impact assessment of a possible urban rail initiative. Report no. ITLR-T17297-005. Prepared by Interfleet Technology Ltd., Pride Parkway, Derby, UK.

10. Bhuyan, S. K. 1982. Cost of quality as a customer perception. *Proceedings of the American Society for Quality Control Annual Congress* 459–464.

11. Zalud, F. H., and Lanc, J. 1972. Automobile reliability: A key to lower overall transport costs. *Proceedings of the 14th International Automobile Technical Congress of FISITA,* London, 6125–6132. Published by the Institution of Mechanical Engineers, London.

9

Civil Engineering Structures and Energy Systems Life Cycle Costing

9.1 Introduction

In recent years, energy conservation has received considerable attention because the escalation of fuel prices has made energy costs an important consideration in the procurement of a wide range of items or systems. In the development and construction of many civil engineering systems and buildings, cost has become an increasingly important issue because past experience indicates that operating and maintenance costs over the long life of a system or building far exceed initial costs. Thus, operating and maintenance costs must be factored into the decision process concerning procurement and construction of civil engineering systems and buildings because it may be more cost effective to take on a higher initial cost in order to obtain lower ownership costs of these items.

The concept of life cycle costing has frequently been used in making procurement and construction decisions concerning energy and civil engineering systems. Over the years, a large number of publications on both these areas have appeared. This chapter presents various important aspects of civil engineering and energy systems life cycle costing.

9.2 Building Life Cycle Cost

In the past, decisions in the building industrial sector during the design phase were made basically by comparing initial capital costs. The main reason for using this approach was its simplicity. Various studies conducted over the years indicate that a building's long-term costs can far outweigh initial capital costs [1,2].

Thus, estimating the life cycle cost of a building at the initial design stage is very important, because past experiences indicate that the earliest decisions tend to establish boundaries to a certain degree for the later ones. According to Khanduri, Bedard, and Alkass [2], around 75–95% of the total life cycle costs of a typical building are locked in by the time its working drawings are prepared. Furthermore, if an estimate of the total life cycle cost is available at an early design stage of a building project, then it is relatively easy to take appropriate cost reduction measures. However, once the project goes into construction, chances to influence the total project cost are reduced quite significantly.

Building life cycle cost is defined by [2]

$$LCC_b = CC + OC + RMC + DC \tag{9.1}$$

where

LCC_b is life cycle cost of a building.

CC is capital cost, which is composed of land and construction costs.

OC is operation cost associated with items such as energy, insurance, and wages.

RMC is repair and maintenance cost.

DC is demolition cost.

9.3 Steel Structure Life Cycle Cost

Life cycle cost of a steel structure is the total cost during its life span. Mathematically, it is expressed as follows [3,4]:

$$LCC_{St} = IC + MC + INC + RC + OC + FC + DC \tag{9.2}$$

where

LCC_{St} is life cycle cost of a steel structure.

IC is initial cost. This includes cost of planning and design; erection cost; cost of preparing the project site, including the cost of the foundation; storage, handling, and receiving costs associated with fabricated pieces and rolled sections; material cost of structural members such as columns, bracings, and beams; fabrication cost, including the material costs of connection elements or components; cost associated with operation of machinery and tools at the construction site; and cost associated with transporting rolled sections to the fabrication shop and transporting the fabricated pieces to the construction site [4].

MC is maintenance cost associated with items such as painting of exposed parts of the steel structure.

INC is inspection cost associated with preventing potentially severe damage to the structure.

RC is repair cost.

OC is operating cost associated with the proper use of the structure for items such as electricity and heating.

FC is probable failure cost. This cost is based on an acceptable probability of failure.

DC is demolishing or dismantling cost.

Past experience indicates that the following main factors influence the life cycle cost of a steel structure [3]:

- structure maintenance policy;
- structure usage;
- cost of the rolled sections used in initial structure construction;
- project site's geographic location;
- expected life of the structure;
- total number of different section types employed in the structure under consideration;
- total number of connections;
- structure importance;
- perimeter of rolled sections in the complete structure;
- currency discount rate; and
- total weight of all rolled sections used in the entire structure.

9.4 Bridge and Waste Treatment Facilities Life Cycle Costs

Life cycle cost analysis is a powerful tool that allows bridge owners or managers to consider the potential consequences of their decisions in present day monetary terms. The life cycle cost of a bridge is expressed by [5,6]

$$LCC_{br} = CONC + INSC + DESC + FAIC + RAMC \qquad (9.3)$$

where
LCC_{br} is bridge life cycle cost.
$CONC$ is construction cost.
$INSC$ is inspection cost.
$DESC$ is design cost.
$FAIC$ is failure cost.
$RAMC$ is repair and maintenance cost.

The life cycle cost of waste treatment facilities is defined by [7]

$$LCC_\omega = CONC + EDIC + OPC + DDC + SRC + WTDC + FEC \qquad (9.4)$$

where
LCC_ω is waste treatment facilities life cycle cost.
$CONC$ is construction cost, which contains the cost of items such as building construction, process equipment, construction management, improvements to land, and site work.

EDIC is engineering, design, and inspection cost.

OPC is operating cost, including the cost of items such as materials, staff, maintenance, peripheral equipment, and utilities.

DDC is decontamination and decommissioning cost. It includes the cost of decontamination and decommissioning (DAD) as well as the cost associated with managing the wastes generated during DAD.

SRC is start-up and readiness review cost and includes the cost of items such as training of personnel, operation and maintenance manuals, initial system testing, and preparation for and performance of contractor readiness reviews.

WTDC is waste transport and disposal cost.

FEC is front-end cost. Usually, this cost includes mostly the cost of activities that are not directly related to producing a new facility but rather are related to regulation. The other important components of the front-end cost are project management cost and cost of preliminary studies such as establishing project definition and developing functional and operational requirements.

9.5 Building Energy Cost Estimation

Over the years, a number of formulas have been developed to estimate the cost of various items concerned with building energy. Some of these formulas are presented next [8,9].

9.5.1 Formula I

This formula is concerned with estimating annual lighting cost and is expressed by [8]

$$LC_a = \frac{(BS)(OT)(EC)}{1000} \tag{9.5}$$

where

LC$_a$ is annual lighting cost.

BS is light bulb size.

OT is light bulb operating period.

EC is electricity cost expressed in dollars per kilowatt hour.

Example 9.1

Assume that a 100 W incandescent light bulb is operated for 9 hours per day for 365 days. Calculate the cost to operte the bulb during the specified period if the electricity cost is $0.4 per kilowatt hour.

By substituting the specified data values into Equation (9.5), we get

$$LC_a = \frac{(100)(9 \times 365)(0.4)}{1000}$$

$$= \$131.4$$

Thus, the total cost to operate the bulb during the given period will be $131.40.

9.5.2 Formula II

This formula is concerned with estimating annual water heating cost and is expressed as follows [8]:

$$WHC_a = \frac{(BTU_a)(FC)}{ER(Btu_p)} \tag{9.6}$$

where
 WHC_a is annual water heating cost.
 BTU_a is annual British thermal units.
 FC is cost per fuel unit.
 ER is efficiency (i.e., the ratio of energy output to energy input).
 Btu_p is British thermal unit per fuel unit.

Example 9.2

Assume that, for a certain manufacturing process, 1,200 gallons of water per hour are needed and water temperature is at 170°F supplied at 60°F for 8 hours per day for 280 days per year. Furthermore, to heat the water, natural gas is burned at 70% efficiency level and its cost is $5 per 1,000 cubic feet (CF). Calculate the annual water heating cost.

By inserting the given data values into Equation (9.6), we get

$$WHC_a = \frac{[(1200)(8.34)(170-60)(8)(280)](\$5/1000\ CF)}{(0.7)(1000\ BTU/CF)}$$

$$= \$17,614.08$$

Thus, the total annual water heating cost will be $17,614.08.

9.5.3 Formula III

This formula is concerned with estimating air filter energy cost. The energy cost, C_e, over the useful life of the filter is expressed by [9,10]

$$C_e = (C_p)(A_q)(FL)(R_f)(K)/(MB_e)(10,000) \tag{9.7}$$

where

C_p is power cost expressed in dollars per kilowatt hour.

A_q is quantity of air to be filtered expressed in cubic feet per minute.

FL is useful life of the air filter expressed in hours.

R_f is filter final resistance expressed in inch water gauge.

K is the constant with value 1.173.

MB_e is motor and blower efficiency.

9.5.4 Formula IV

This formula is concerned with estimating the cost of heat exchangers, and is expressed by [9,11]

$$C_{he} = \theta (SA)^n \tag{9.8}$$

where

C_{he} is procurement cost, free-on-board (F.O.B) factory.

SA is heat exchanger surface area expressed in square feet.

θ and n are constants (their tabulated values are available in references 9 and 11).

9.5.5 Formula V

This formula is concerned with estimating operational equipment energy consumption cost and is expressed by [12]

$$C_{oe} = (P_a)(EP_C)(OH)(OE) \tag{9.9}$$

where

C_{oe} is total energy consumption cost of operational equipment.

P_a is average electrical power rating.

EP_C is electrical power cost.

OH is total number of annual operating hours.

OE is total number of pieces of operational equipment.

9.6 Appliance Life Cycle Costing

There are many different types of appliances—for example, refrigerators, ranges and ovens, freezers, gas dryers, washing machines, electric dryers, and room air conditioners. Their life cycle costs can be estimated by using the following equation [9,13]:

$$LCC_a = AQC + \sum_{i=1}^{K} EC_i \left[\frac{FC(1+fr)^i}{(1+dr)^i} \right] \tag{9.10}$$

where

LCC_a is appliance life cycle cost.

AQC is appliance acquisition cost expressed in dollars.

K is appliance useful life expressed in years.

EC_i is energy consumption of year i expressed in British thermal units (BTUs).

FC is annual fuel cost expressed in constant dollars per million BTUs.

fr is annual fuel escalation rate (%) expressed in constant dollars.

dr is discount rate (%) expressed in constant dollars.

In the case of yearly constant energy consumption, EC, and the fuel escalation rate, fr, over appliance useful life, Equation (9.10) simplifies to

$$LCC_a = AQC + (EC)(FC)\sum_{i=1}^{K} \frac{(1+fr)^i}{(1+dr)^i} \tag{9.11}$$

Past experience indicates that acquisition cost for items such as refrigerators, electric ranges, and room air conditioners accounts for roughly 41, 38, and 59% of their life cycle costs, respectively [12].

9.7 Energy System Life Cycle Cost Estimation Model

This model was developed by the Center for Building Technology of the National Bureau of Standards for the U.S. Department of Energy [13]. The model takes into consideration all relevant costs over time of a building facility's design, materials, operation, systems, and components. More specifically, it includes items such as initial investment cost, operation and maintenance cost, future replacement cost, and salvage and resale value.

Thus, the energy system life cycle cost is expressed by [12,13]

$$LCC_{es} = EC_{pv} + IC_{pv} + SV_{pv} + NFOMC_{pv} + NRC_{pv} + RC_{pv} \tag{9.12}$$

where

LCC_{es} is present value of the energy system life cycle cost.

EC_{pv} is present value of the energy cost.

IC_{pv} is present value of the investment cost.

SV_{pv} is present value of salvage.

$NFOMC_{pv}$ is present value of the annually recurring nonfuel operation and maintenance cost.

NRC_{pv} is present value of the nonrecurring nonfuel operation and maintenance cost.

Additional information on this model is available in references 12 and 13.

9.8 Motor, Pump, and Circuit-Breaker Life Cycle Costs

This section presents mathematical models to estimate life cycle cost of a motor, a pump, and a circuit breaker.

9.8.1 Motor Life Cycle Cost Estimation Model

This model is concerned with estimating the life cycle cost of an electric motor, which is expressed by [9,14]

$$LCC_m = C_{ma} + C_{mo} \qquad (9.13)$$

where
  LCC_m is motor life cycle cost.
  C_{ma} is motor acquisition cost.
  C_{mo} is motor operating cost.

Note that in Equation (9.13), the motor maintenance cost is assumed negligible. Using Dhillon [9], the present value of the motor operating cost, C_{moj}, for year j may be expressed as follows:

$$PV_j = C_{moj} \left[\frac{1}{1+i} \right]^j \qquad (9.14)$$

where
  PV_j is present value of the motor operating cost, C_{moj}, for year j.
  i is interest rate.

If the motor operational life is m years, then the present value of the motor total operating cost is expressed by

$$C_{mot} = C_{mo1} \left(\frac{1}{1+i} \right) + C_{mo2} \left(\frac{1}{1+i} \right)^2 + C_{mo3} \left(\frac{1}{1+i} \right)^3 + \cdots + C_{mom} \left(\frac{1}{1+i} \right)^m \qquad (9.15)$$

where
  C_{mot} is present value of the total operating cost of the motor.
  C_{moj} is motor operating cost in year j for $j = 1, 2, 3,..., m$.

The yearly operating cost of the motor can be calculated by using the following equation [8,14]:

$$C_{mo} = \frac{(OH)(0.746)(MS)(C_e)}{EFF} \qquad (9.16)$$

where

C_{mo} is motor operating cost per year expressed in dollars.

OH is annual motor operating hours.

MS is motor size expressed in horsepower.

C_e is cost of electricity expressed in dollars per kilowatt hour.

EFF is motor efficiency.

Example 9.3

Assume that a 30-horsepower electric motor is operated for 3,000 hours annually. The cost of electrical energy is $0.2 per kilowatt hour. Calculate the annual cost to operate the motor if motor efficiency is 95%.

By substituting the given data values into Equation (9.16), we get

$$C_{mo} = \frac{(3000)(0.746)(30)(0.2)}{0.95}$$

$$= \$14,137.74$$

Thus, the annual cost to operate the motor will be $14,137.74.

9.8.2 Pump Life Cycle Cost Estimation Model

This model is concerned with estimating the life cycle cost of a pump, which is expressed by [15,16]

$$LCC_p = IC + EC + IAC + DC + DTC + OC + MRC + ENC \qquad (9.17)$$

where

LCC_p is pump life cycle cost.

IC is pump initial cost, including the cost of items such as pump, pipe, system, and auxiliary.

EC is pump energy cost associated with various aspects of pump system operation.

IAC is pump installation and commissioning cost, including the cost of training.

DC is pump decommissioning or disposal cost, which also includes the cost associated with restoration of the local environment and disposal of auxiliary services.

DTC is pump downtime cost associated with the production losses.

OC is pump operation cost, which is basically the labor cost of normal pump system supervision.

MRC is pump maintenance and repair cost.

ENC is pump environmental cost associated with contamination from pumped liquid.

Each of these eight costs is described in detail in reference 16.

The pump energy cost, *EC*, may be calculated by using the following formula [8]:

$$EC = \frac{(PS)(PHS)(AOP)(C_e)}{(5300)\theta_p \theta_m} \tag{9.18}$$

where
PS is pump size expressed in gallons per minute (GPM).
PHS is pump head size expressed in feet.
AOP is pump annual operational period expressed in hours.
C_e is cost of electricity expressed in dollars per kilowatt hour.
θ_p is pump efficiency.
θ_m is motor efficiency.

Example 9.4

Assume that an 800 GPM pump with a total head size of 10 feet is operated for 1,500 hours per year. The pump and motor efficiency are 70 and 90%, respectively. Calculate the annual cost to operate the pump if the cost of electricity is $0.3 per kilowatt hour.

By substituting the specified data values into Equation (9.18), we get

$$EC = \frac{(800)(10)(1\,500)(0.3)}{(5300)(0.7)(0.9)}$$

$$= \$1,078.17$$

Thus, the annual cost to operate the pump will be $1,078.17.

9.8.3 Circuit-Breaker Life Cycle Cost Estimation Model

This model is concerned with estimating the life cycle cost of a high-voltage circuit breaker. This cost is expressed by [9,17]

$$LCC_{cb} = CFC + CMC + UC \tag{9.19}$$

where
LCC_{cb} is life cycle cost of the high-voltage circuit breaker.
CFC is high-voltage circuit breaker fixed cost.
CMC is high-voltage circuit breaker maintenance cost.
UC is cost associated with the unavailability of power transmission and distribution systems.

Problems

1. Write an essay on building life cycle costing.
2. Write an equation for estimating life cycle cost of a building.
3. Write an equation that can be used to estimate life cost of a steel structure.

4. Write at least 10 factors that influence the life cycle cost of a steel structure.
5. Write an equation that can be used to estimate life cycle cost of a bridge.
6. Assume that a 60 W incandescent light bulb is operated for 6 hours per day for 365 days. Calculate the cost to operate the bulb during the specified period if the electricity cost is $0.3 per kilowatt hour.
7. Write formulas for estimating the cost of (1) heat exchangers, and (2) filter energy.
8. Assume that for a certain manufacturing process, 1,000 gallons of water per hour is required and water temperature is at 150°F supplied at 70°F for 6 hours per day for 250 days per year. Furthermore, to heat the water, natural gas is burned at 60% efficiency level and its cost is $6 per 1,000 CF. Calculate the annual water heating cost.
9. Write an equation that can be used to estimate life cycle costs of refrigerators and washing machines.
10. Assume that a 20-horsepower electric motor is operated for 2,000 hours annually. The cost of electrical energy is $0.3 per kilowatt hour. Calculate the annual cost to operate the motor if the motor efficiency is 90%.

References

1. Flanagan, R., Norman, G., Meadows, J., and Robinson, G. 1989. *Life cycle costing, theory and practice.* London: BSP Professional Books.
2. Khanduri, A. C., Bedard, C., and Alkass, S. 1983. Life cycle costing of office buildings at the preliminary design stage. *Proceedings of the 5th International Conference on Civil and Structural Engineering Computing* 1–8.
3. Sarma, K. C., and Adeli, H. 2002. Life cycle cost optimization of steel structures. *International Journal for Numerical Methods in Engineering* 55:1451–1462.
4. Sarma, K. C., and Adeli, H. 2000. Fuzzy discrete multi-criteria cost optimization of steel structures. *Journal of Structural Engineering (ASCE)* 126 (11): 1339–1347.
5. Rafiq, M. I., Chryssanthopoulos, M., and Onoufriou, T. 2005. Comparison of bridge management strategies using life cycle cost analysis. *Proceedings of the 5th International Conference on Bridge Management, Inspections, Maintenance, Assessment, and Repair* 578–586.
6. Frangopol, D. M., Estes, A. C., Augusti, G., and Ciampoli, M. 1997. Optimal bridge management based on lifetime reliability and life cycle costs. *Proceedings of the International Workshop on Optimal Performance of Civil Infrastructure Systems* 98–115.
7. Sivill, T. E., Stoddard, D. N., Smith, T. H., and Roesener, W. S. 1993. Use of life cycle cost estimates in the evaluation of proposed waste-treatment facilities. *Proceedings of the Technology and Programs for Radioactive Waste Management and Environmental Restoration Conference* 1797–1801.
8. Brown, R. J., and Yanuck, R. R. 1980. *Life cycle costing: A practical guide for energy managers.* Atlanta, GA: Fairmont Press.

9. Dhillon, B. S. 1989. *Life cycle costing: Techniques, models and applications.* New York: Gordon and Breach Science Publishers.
10. Avery, A. H. 1977. Life cycle costing of high-efficiency air filters. *Plant Engineering* September: 80–83.
11. Kumana, J. D. 1984. Cost update on specialty heat exchangers. *Chemical Engineering* June: 169–172.
12. Earles, M. E. 1981. *Factors, formulas, and structures for life cycle costing.* Concord, MA: Eddins–Earles.
13. National Bureau of Standards. 1980. *Life cycle cost manual for the Federal Energy Management Program.* National Bureau of Standards handbook 135. U.S. Department of Commerce, Washington, D.C.
14. Ganapathy, V. 1983. Life cycle costing applied to motor selection. *Process Engineering* July: 51–52.
15. De Boer, G., and Greidanus, D. 2006. Utilization of customized hydraulics to elongate pump life and lower cycle costs. *Proceedings of the Institution of Mechanical Engineers 9th European Fluid Machinery Congress on Applying the Latest Technology to New and Existing Process Equipment* 95–102.
16. Hydraulic Institute, Office of Industrial Technology. 2001. Pump life cycle costs: A guide to LCC analysis for pumping systems. Report no. DOE/GO-102001-1190, U.S. Department of Energy, Washington, D.C.
17. Heising, C. R. 1979. Reliability and maintenance data needed for high-voltage circuit breakers when making life cycle cost studies. *Proceedings of the International Reliability Conference for the Electric Power Industry* 103–108.

10

Miscellaneous Cost Estimation Models

10.1 Introduction

Over the years, a large number of cost estimation models have been developed in diverse areas ranging from software engineering to telecommunication engineering. A cost estimation model may be described simply as an approach, based on programmatic and technical parameters, to calculating costs under consideration. More specifically, some of the possible dimensions of a cost estimation model include the elements of cost, time, and cost breakdown structure.

Many desirable features are associated with a cost estimation model; in designing such a model, the main factors that should be considered are feasibility of data requirements; operation ease; cost to develop, operate, and alter; capability for sensitivity analyses; speed to set up, operate, and change; inclusiveness and authoritativeness; and tolerance of input errors [1,2].

There are various types of cost estimation models: capital cost estimation models, operation and maintenance cost estimation models, life cycle cost estimation models, and so on. This chapter presents a number of models that were not covered in previous chapters. They can also be used to estimate various types of costs for performing life cycle cost analysis directly or indirectly.

10.2 Plant Cost Estimation Model

This model was developed by Cran to estimate plant cost in the chemical industry [3]. The total plant cost is expressed by [3,4]

$$TPC = C_d + C_i$$
$$= C_d + (C_d)(C_{if})$$

(10.1)

where
TPC is total plant cost expressed in dollars.
C_d is direct cost expressed in dollars.
C_i is indirect cost expressed in dollars
C_{if} is indirect cost factor.

The direct cost, C_d, is defined by

$$C_d = (C_{is})(C_{di}) + (C_e)(C_{dp}) \tag{10.2}$$

where
C_{is} is cost of instruments.
C_{di} is direct cost factor associated with instruments.
C_e is cost of equipment.
C_{dp} is direct cost factor associated with the plant.

The following are the mean values for plant direct cost and instrument direct cost factors:

- 2.16 (for C_{dp})
- 2.50 (for C_{di})

The value of the indirect cost factor, C_{if}, can be estimated by using the following equation:

$$C_{if} = 1.36 - (0.073) ln C_d \tag{10.3}$$

Additional information on the model is available in Cran [3] and Ward [4].

10.3 Reliability Acquisition Cost Estimation Model

This model can be used to estimate reliability acquisition cost when state-of-the-art system acquisition cost and reliability improvement ratio compared to state of the art are known. The reliability acquisition cost is expressed by [5]

$$C_{ra} = (0.2) (AC_{sa}) ln \alpha \tag{10.4}$$

where
C_{ra} is reliability acquisition cost.
AC_{sa} is state-of-the-art system acquisition cost.
α is reliability improvement ratio compared to state of the art.

Additional information on the model is available in Winlund [5].

10.4 Development Cost Estimation Model

This model is concerned with estimating development cost by considering the reliability factor. Thus, the development cost is expressed by [6,7]

$$DC_r = C_{ir} + C_{dr} \tag{10.5}$$

where

DC_r is development cost, considering the reliability.
C_{ir} is basic cost, independent of reliability.
C_{dr} is cost, dependent on reliability (i.e., reliability-related cost).

The cost dependent on reliability, C_{dr}, is defined by

$$C_{dr} = C_s \left[\frac{MTBF_i}{MTBF_s} \right]^{\theta} \tag{10.6}$$

where

C_s is "standard" cost to develop item, equipment, or system having "standard" or current reliability.
$MTBF_i$ is item, equipment, or system mean time between failures with improved design.
$MTBF_s$ is item, equipment, or system mean time between failures with standard design.
θ is a constant whose value is to be estimated from empirical studies.

Let us now assume that the reliability of the standard design, $R_s(t)$, and the reliability of the improved design, $R_i(t)$, are respectively expressed by [8]

$$R_s(t) = e^{-\left(\frac{t}{MTBF_s}\right)} \tag{10.7}$$

and

$$R_i(t) = e^{-\left(\frac{t}{MTBF_i}\right)} \tag{10.8}$$

where

t is time.
$R_s(t)$ is standard design reliability at time t.
$R_i(t)$ is improved design reliability at time t.

By taking natural logarithms of Equation (10.7) and (10.8) and then rearranging them, we get, respectively,

$$MTBF_s = -\left[\frac{t}{\ln R_s(t)}\right] \tag{10.9}$$

and

$$MTBF_i = -\left[\frac{t}{\ln R_i(t)}\right] \tag{10.10}$$

By substituting Equations (10.9) and (10.10) into Equation (10.6) and then substituting the resulting equation into Equation (10.5), we get

$$DC_r = C_{ir} + C_s\left[\frac{\ln R_s(t)}{\ln R_i(t)}\right]^{\theta} \tag{10.11}$$

Note that the preceding equation makes use of time-dependent reliabilities of standard and improved item, equipment, or system designs instead of mean time between failures (i.e., $MTBF_s$ and $MTBF_i$) as in the case of Equation (10.6). Additional information on the model is available in Hevesh [6] and Carhart and Herd [7].

10.5 Program Error Cost Estimation Model

This model is concerned with estimating the cost of program errors in a program and is expressed by [8]

$$PEC = \sum_{j=1}^{m}(C_{oj} + C_{cj}) \tag{10.12}$$

where
   PEC is total cost of errors in a program.
   m is total number of errors in a program.
   C_{oj} is cost associated with the occurrence of error j.
   C_{cj} is cost associated with correcting error j.

Note that the cost elements associated with the error occurrence cost are lost equipment time cost, wasted manpower hours cost, etc. Similarly, the cost of correcting the error includes components such as equipment cost, supply cost, and manpower cost. Additional information on the model is available in Sontz [8].

10.6 Cooling Tower Cost Estimation Model

This model is concerned with estimating cooling tower cost using operating data. This cost is expressed by [9]

$$C_t = \frac{L}{(Z)(X) - 586 + (39.2)(R)} \tag{10.13}$$

$$Z = \frac{279}{(0.0335)(85 - T_{wb})^{1.143} + 1} \tag{10.14}$$

$$R = T_{hw} - T_{cw} \tag{10.15}$$

$$X = T_{cw} - T_{wb} \tag{10.16}$$

where
C_t is cost of a cooling tower expressed in dollars.
L is total heat load expressed in BTUs per hour.
X is the temperature approach.
R is cooling range.
T_{wb} is wet bulb temperature expressed in degrees Fahrenheit.
T_{hw} is hot water temperature expressed in degrees Fahrenheit.
T_{cw} is cooled water temperature expressed in degrees Fahrenheit.

Additional information on the model is available in Zanker [9].

Example 10.1
Calculate the cost of a cooling tower, if the following data values are given:

$T_{hw} = 120°F;$
$T_{cw} = 80°F;$
$T_{wb} = 60°F;$ and
$L = 300$ million BTUs per hour.

By substituting the given data values into Equations (10.13)–(10.16), we get

$$R = 120 - 80 = 40°F$$
$$X = 80 - 60 = 20°F$$
$$Z = \frac{279}{(0.0335)(85 - 60)^{1.143} + 1} = 119.89$$
$$C_t = \frac{300,000,000}{(119.89)(20) - 586 + (39.2)(40)} = \$88,760.64$$

Thus, the cooling tower cost is \$88,760.64.

10.7 Storage Tank Cost Estimation Model

This model is concerned with estimating the cost of storage tanks and is expressed by [10]

$$C_{st} = C_b(\lambda) \qquad (10.17)$$

where
C_{st} is cost of a storage tank.
C_b is base cost, in carbon steel, expressed in dollars.
λ is the material-of-construction factor.

The base cost in carbon steel for field-erected tanks (cone roofs and flat bottoms) is expressed by

$$C_b = \exp\left[\theta_1 - \theta_2 \ln V + \theta_3 (\ln V)^2\right] \qquad (10.18)$$

where
C_b is base cost in carbon steel for field-erected tanks.
θ_j is the jth constant for $j = 1$ ($\theta_1 = 9.369$), $j = 2$ ($\theta_2 = 0.1045$), and $j = 3$ ($\theta_3 = 0.045355$).
V is tank volume in cubic meters (80 m³ $\leq V \leq$ 45,000 m³).

Similarly, the base cost in carbon steel for shop-fabricated tanks (cone roofs and flat bottoms) is expressed by

$$C_b = \exp\left[\alpha_1 + \alpha_2 \ln V - \alpha_3 (\ln V)^2\right] \qquad (10.19)$$

where
C_b is base cost in carbon steel for shop-fabricated tanks.
α_j is the jth constant for $j = 1$ ($\alpha_1 = 7.994$), $j = 2$ ($\alpha_2 = 0.6637$), and $j = 3$ ($\alpha_3 = 0.063088$).
V is tank volume in cubic meters (5 m³ $\leq V \leq$ 80 m³)

The values of λ for construction materials such as stainless steel 304, stainless steel 316, stainless steel 347, aluminum, copper, nickel, titanium, and monel are 2.4, 2.7, 3, 2.7, 2.3, 3.5, 11.0, and 3.3, respectively. Additional information on the model is available in Corripio, Chrien, and Evans [10].

10.8 Pressure Vessel Cost Estimation Model

This model is concerned with estimating the cost of pressure vessels. The total cost is expressed by [11]

$$PVC = (\theta)C_v + C_p \qquad (10.20)$$

where

PVC is total cost of a pressure vessel expressed in dollars.
θ is construction material cost factor.
C_v is base vessel cost in carbon steel expressed in dollars.
C_p is cost of platform and ladders expressed in dollars.

For horizontal vessels, C_v and C_p are expressed by Equations (10.21) and (10.22), respectively:

$$C_v = \exp\left[L_1 - L_2\, ln\, W + L_3\, (ln\, W)^2\right] \tag{10.21}$$

where

L_j is the jth constant for $j = 1$ ($L_1 = 8.114$), $j = 2$ ($L_2 = 0.16449$), and $j = 3$ ($L_3 = 0.04333$).
W is carbon steel shell weight in kilograms (369 kg $\leq W \leq$ 415,000 kg).

$$C_p = M_1\, (D^{M_2}) \tag{10.22}$$

where

M_j is the jth constant for $j = 1$ ($M_1 = 1288.3$) and $j = 2$ ($M_2 = 0.20294$).
D is inside diameter of platform and ladders in meters (0.92 m $\leq D \leq$ 3.66 m).

Similarly, for vertical vessels, C_v and C_p are expressed by Equations (10.23) and (10.24), respectively:

$$C_v = \exp\left[N_1 - N_2\, ln\, W + N_3\, (ln\, W)^2\right] \tag{10.23}$$

where

N_j is the jth constant for $j = 1$ ($N_1 = 8.6$), $j = 2$ ($N_2 = 0.21651$), and $j = 3$ ($N_3 = 0.04576$).
W is carbon steel shell weight in kilograms (2210 kg $\leq W \leq$ 103,000 kg).

$$C_p = n_1\, D^{n_2}\, (TL)^{n_3} \tag{10.24}$$

where

n_j is the jth constant for $j = 1$ ($n_1 = 1017$), $j = 2$ ($n_2 = 0.73960$), and $j = 3$ ($n_3 = 0.70684$).
D is inside diameter of platform and ladders in meters (1.83 m $\leq D \leq$ 3.05 m).

The values of θ for construction materials such as stainless steel 316, stainless steel 304, titanium, nickel 200, monel 400, incoloy 825, and inconel 600 are 2.1, 1.7, 7.7, 5.4, 3.6, 3.7, and 3.9, respectively. Additional information on the model is available in Mulet, Corripio, and Evans. [11].

10.9 New Aircraft System Spares Cost Estimation Model

This model is concerned with estimating spares cost for the new aircraft system. The model uses the spares cost for an existing aircraft system and adjusts it by a comparison factor reflecting the differences in system cost, reliability, hardware complexity, and repairability. The new aircraft system's spares cost is defined by [12]

$$C_{na} = \gamma\, C_{ea} \tag{10.25}$$

where
  C_{na} is new aircraft system spares cost.
  θ is the comparison factor expressed in terms of operational and support
      parameters.
  C_{ea} is existing aircraft system spares cost.

The comparison factor, θ, is expressed by

$$\theta = \beta q \left[\frac{MI_n}{MI_e} \right] \tag{10.26}$$

where
  q is quantifier of the cost impact associated with a shift in the classifica-
      tion of spares from "base repaired" to "depot repaired" or vice versa
      between the two aircraft systems under consideration. Note that the
      value of this quantifier is equal to unity when the change in the ratio
      of the two classifications is zero.
  MI_n is new aircraft system's calculated (estimated) maintenance index
      expressed as maintenance man-hours per flying hour.
  MI_e is existing aircraft system's established maintenance index expressed
      as maintenance man-hours per flying hour.

The symbol β in Equation (10.26) is expressed by

$$\beta = f_1 \left[\frac{C_{n1}}{C_{e1}} \right] + f_2 \left[\frac{C_{n2}}{C_{e2}} \right] + f_3 \left[\frac{C_{n3}}{C_{e3}} \right] \tag{10.27}$$

where
  C_{nj} is jth segment of the "fly-away" cost for the new aircraft system for $j = 1$
      (airframe), $j = 2$ (propulsion), and $j = 3$ (equipment).
  C_{ej} is jth segment of the "fly-away" cost for the existing aircraft system for
      $j = 1$ (airframe), $j = 2$ (propulsion), and $j = 3$ (equipment).

f_j is jth fraction of the total investment spares "lay-in" value calculated for existing systems for $j = 1$ (airframe related), $j = 2$ (propulsion related), and $j = 3$ (equipment related).

Additional information on the model is available in Tyszkiewicz [12].

10.10 Satellite Procurement Cost Estimation Model

This model is concerned with estimating the procurement cost of satellites in 1974 dollars. The satellite procurement cost is expressed by [2,13]

$$C_s = \left[\lambda_1 W_s^{-\lambda_2} \right] W_s \tag{10.28}$$

where
  C_s is satellite procurement cost.
  λ_j is the jth constant for $j = 1$ ($\lambda_1 = 1{,}970{,}300$) and $j = 2$ ($\lambda_2 = 0.592$).
  W_s is the satellite's total weight.

Additional information on the model is available in Hadfield [13].

10.11 Single-Satellite System Launch Cost Estimation Model

This model is concerned with estimating the total launch cost of a single-satellite system (circular orbits). The launch cost is defined by [13]

$$LC_s = \beta_1 (W_s)^{\beta_2} \left[\frac{OA}{\beta_3} + f_p + 1 \right] \tag{10.29}$$

where
  LC_s is launch cost of a single-satellite system, expressed in millions of 1974 dollars.
  β_j is the jth constant for $j = 1$ ($\beta_1 = 0.026$), $j = 2$ ($\beta_2 = 2/3$), and $j = 3$ ($\beta_3 = 8{,}000$).
  W_s is total satellite weight expressed in pounds.
  OA is orbit altitude or apogee expressed in statute miles.
  f_p is a factor measuring the satellite payload sophistication.

Additional information on the model is available in Hadfield [13].

10.12 Tank Gun System Life Cycle Cost Estimation Model

This model is concerned with estimating tank gun life cycle cost by decomposing it into three major components: research and development cost, investment cost, and operating and support cost. The life cycle cost is expressed by [1,2]

$$LCC_{tg} = RDC_{tg} + IC_{tg} + OS_{tg} \tag{10.30}$$

where
LCC_{tg} is tank gun system life cycle cost.
RDC_{tg} is tank gun system research and development cost.
IC_{tg} is tank gun system investment cost.
OS_{tg} is tank gun system operating and support cost.

The tank gun system research and development cost, RDC_{tg}, is expressed by

$$RDC_{tg} = \sum_{j=1}^{10} RDC_{tgj} \tag{10.31}$$

where
RDC_{tgj} is cost component j of the tank gun system research and development cost for
$j = 1$ (tooling cost)
$j = 2$ (facilities cost)
$j = 3$ (development engineering cost)
$j = 4$ (system project management cost)
$j = 5$ (prototype manufacturing cost)
$j = 6$ (system test and evaluation cost)
$j = 7$ (training cost)
$j = 8$ (producibility engineering and planning cost)
$j = 9$ (data cost)
$j = 10$ (other cost)

The tank gun system investment cost, IC_{tg}, is expressed by

$$IC_{tg} = \sum_{j=1}^{11} IC_{tgj} \tag{10.32}$$

where
IC_{tgj} is cost component j of the tank gun system investment cost for
$j = 1$ (training cost)
$j = 2$ (production cost)

$j = 3$ (data cost)
$j = 4$ (nonrecurring investment cost)
$j = 5$ (system project management cost)
$j = 6$ (initial spares and repair parts cost)
$j = 7$ (engineering changes cost)
$j = 8$ (transportation cost)
$j = 9$ (system test evaluation cost)
$j = 10$ (operational and site activation cost)
$j = 11$ (other cost)

The tank gun system operation and support cost, OS_{tg}, is expressed by

$$OS_{tg} = \sum_{j=1}^{6} OS_{tgj} \tag{10.33}$$

where
OS_{tgj} is the cost component j of the tank gun system operating and support cost for
$j = 1$ (consumption cost)
$j = 2$ (modification material cost)
$j = 3$ (military personnel cost)
$j = 4$ (depot maintenance cost)
$j = 5$ (other direct support operations cost)
$j = 6$ (indirect support and operations cost)

Additional information on the model is available in Earles [1] and Dhillon [2].

10.13 Weather Radar Life Cycle Cost Estimation Model

This model is concerned with estimating the life cycle cost of weather radar. This cost is expressed by [1]

$$WRLCC = SDC + VC + AC + OMSC \tag{10.34}$$

where
$WRLCC$ is weather radar life cycle cost.
SDC is weather radar system definition cost.
VC is weather radar validation cost.
AC is weather radar acquisition cost.
$OMSC$ is weather radar operation, maintenance, and support cost.

The weather radar system definition cost, SDC, is expressed by

$$SDC = \sum_{i=1}^{2} SDC_i \qquad (10.35)$$

where SDC_i is the ith cost component of the weather radar system definition cost for $i = 1$ (program management cost) and $i = 2$ (cost for each bidder).

The weather radar system validation cost, VC, is expressed by

$$VC = \sum_{i=1}^{15} VC_i \qquad (10.36)$$

where
  VC_i is the ith cost component of the weather radar validation cost for
  $i = 1$ (engineering design and development cost)
  $i = 2$ (fabrication and manufacturing development cost)
  $i = 3$ (validation hardware cost)
  $i = 4$ (software system design and development cost)
  $i = 5$ (logistics planning and support cost)
  $i = 6$ (development test and test support cost)
  $i = 7$ (validation test site preparation cost)
  $i = 8$ (documentation cost)
  $i = 9$ (manual development cost)
  $i = 10$ (support and test equipment cost)
  $i = 11$ (training development and equipment cost)
  $i = 12$ (government-furnished equipment and facilities cost)
  $i = 13$ (transportation of equipment to test site cost)
  $i = 14$ (program management cost)
  $i = 15$ (general and administration cost)

The weather radar acquisition cost, AC, is expressed by

$$AC = \sum_{i=1}^{18} AC_i \qquad (10.37)$$

where
  AC_i is ith cost component of the weather radar acquisition cost for
  $i = 1$ (software and firmware-manufacturing-related cost)
  $i = 2$ (software and firmware-depot-related cost)
  $i = 3$ (software and firmware-on-site-related costs)
  $i = 4$ (initial training cost)
  $i = 5$ (vendor warranty for first year cost)
  $i = 6$ (test and support equipment cost)
  $i = 7$ (initial spares cost)

$i = 8$ (test and evaluation cost)
$i = 9$ (data and documentation cost)
$i = 10$ (site preparation cost)
$i = 11$ (system installation and checkout cost)
$i = 12$ (site decommissioning cost)
$i = 13$ (land acquisition cost)
$i = 14$ (government printing cost)
$i = 15$ (manual binding and delivery cost)
$i = 16$ (government-furnished equipment cost)
$i = 17$ (program management cost)
$i = 18$ (general and administration overhead cost)

The weather radar operation, maintenance, and support cost, OMSC, is expressed by

$$OMSC = \sum_{i=1}^{13} OMSC_i \qquad (10.38)$$

where
   $OMSC_i$ is ith cost component of the weather radar operation, maintenance, and support cost for
$i = 1$ (operating personnel cost)
$i = 2$ (electric power cost)
$i = 3$ (communications facilities cost)
$i = 4$ (occupying and housekeeping cost)
$i = 5$ (consumables cost)
$i = 6$ (dedicated maintenance personnel cost)
$i = 7$ (other maintenance-preventive and corrective cost)
$i = 8$ (recurring spares cost)
$i = 9$ (logistics and logistics support cost)
$i = 10$ (other maintenance-test and support cost)
$i = 11$ (equipment rental and housekeeping cost)
$i = 12$ (maintenance by contractor cost)
$i = 13$ (recurring training cost)

Additional information on the model is available in Earles [1].

Problems

   1. What is a cost estimation model?
   2. Write an essay on cost estimation models.
   3. Discuss the desirable features of a cost estimation model.

4. Define two main elements of a plant cost estimation model.
5. Define the following two models: (1) reliability acquisition cost estimation model, and (2) development cost estimation model.
6. Define program error cost estimation model.
7. If the following data values are known, estimate the cost of a cooling tower by using Equation (10.13):
   - $L = 400$ million BTUs per hour
   - $T_{hw} = 130°F$
   - $T_{cw} = 90°F$
   - $T_{wb} = 65°F$
8. Define satellite acquisition cost estimation model.
9. Discuss the following two items: (1) weather radar life cycle cost and (2) tank gun system life cycle cost
10. Define the following two models: (1) storage tank cost estimation model and (2) pressure vessel cost estimation model.

References

1. Earles, M. 1981. *Factors, formulas, and structures for life cycle costing.* Concord, MA: Eddins–Earles.
2. Dhillon, B. S. 1989. *Life cycle costing: Techniques, models, and applications.* New York: Gordon and Breach Science Publishers.
3. Cran, J. 1981. Improved factored method gives better preliminary cost estimates. *Chemical Engineering* April: 65–79.
4. Ward, T. J. 1986. Cost-estimating methods. In *Design of Equipment, vol. 1: Plant design and cost estimating,* ed. J. Beckman, 12–21. New York: American Institute of Chemical Engineers.
5. Winlund, E. S. 1965. Cost-effective analysis for optimal reliability and maintainability. *Proceedings of the Annual National Symposium on Reliability and Quality Control* 107–114.
6. Hevesh, A. H. 1969. Cost of reliability improvement. *Proceedings of the Annual Symposium on Reliability* 54–61.
7. Carhart, R. R., and Herd, G. R. 1957. A simple cost model for optimizing reliability. In Reliability of military electronic equipment, a report by Advisory Group on Reliability of Electronic Equipment (AGREE), Office of the Assistant Secretary of Defense (Research and Engineering), Department of Defense, Washington, D.C., 64–74.
8. Sontz, C. 1973. Quality assurance for the data processing industry. *Proceedings of the Annual Reliability and Maintainability Symposium* 136–148.
9. Zanker, A. 1972. Estimating cooling tower costs from operating data. *Chemical Engineering* June: 118–120.
10. Corripio, A. B., Chrien, K. S., and Evans, L. B. 1982. Estimate costs of heat exchangers and storage tanks via correlations. *Chemical Engineering* January: 125–127.

11. Mulet, A., Corripio, A. B., and Evans, L. B. 1981. Estimate costs of pressure vessels via correlations. *Chemical Engineering* October: 145–150.
12. Tyszkiewicz, A. M. 1983. A comparative cost model for aircraft investment spares. *Proceedings of the Annual Reliability and Maintainability Symposium* 376–382.
13. Hadfield, B. B. 1974. Satellite-systems cost estimation. *IEEE Transactions on Communications* 22:1540–1547.

11

Introduction to Engineering Reliability and Maintainability

11.1 Introduction

Reliability may be described simply as the probability that an item or system will perform its mission satisfactorily for the desired period when used according to designed conditions. The history of the reliability field may be traced back to the early 1930s, when probability-related concepts were applied to various problems associated with electric power generation [1–4]. During World War II, Germans applied various basic reliability concepts to improve reliability of their rockets (i.e., V1 and V2 rockets). A detailed history of the reliability field is available in Dhillon [5]. Today, reliability is a well-established discipline and has branched out into many specialized areas [5,6].

Maintainability may be described as the aspects of equipment or an item that improve repairability and serviceability, increase cost effectiveness of maintenance, and ensure that the equipment or item satisfies the requirements for its intended application. The roots of the maintainability history may be traced back to 1901 in the Army Signal Corps contract for development of the Wright Brothers' airplane, which stated that the aircraft "should be simple to operate and maintain." However, in the modern context, the beginning of the maintainability discipline may be traced back to the period between World War II and the early 1950s [7,8]. During this period, the U.S. Department of Defense conducted various studies that indicated startling results concerning the state of reliability and maintainability of equipment used by the three services [8–10].

Needless to say, today reliability and maintainability are well-established disciplines and, over the years, a vast amount of literature on both the topics has appeared [11,12]. This chapter presents various fundamental aspects of reliability and maintainability considered useful for direct or indirect applications in life cycle costing.

11.2 Reliability and Maintainability Definitions

Some of the commonly used terms and definitions in the area of reliability and maintainability follow [13–16]:

- *Reliability* is the probability that an item will perform its assigned mission satisfactorily for the desired period when used according to specified conditions.
- *Maintainability* is the probability that a failed item will be restored to its satisfactory working state within a stated total downtime, when maintenance activity is started according to specified conditions.
- *Failure* is the inability of an item to perform its specified function within defined guidelines.
- *Downtime* is the time during which the item or product is not in a condition to perform its stated mission or function.
- *Availability* is the probability that an item or equipment will be available for service when required.
- *Redundancy* is the existence of more than one means for performing a stated function.
- *Useful life* is the length of time an item or piece of equipment functions within an acceptable level of failure rate.
- *Maintenance* is all scheduled and unscheduled actions necessary to keep an item or piece of equipment in a serviceable state or restoring it to serviceability. It includes items such as inspection, testing, repair, modification, and servicing.
- *Mission time* is the time during which the item or piece of equipment is carrying out its stated mission.

11.3 Bathtub Hazard Rate Curve

The curve shown in Figure 11.1 is widely used to describe failure rate of various types of engineering items. As shown in the figure, the bathtub hazard rate curve is divided into three regions: region I (burn-in period), region II (useful life period), and region III (wear-out period).

- During the burn-in period, hazard rate (i.e., time-dependent failure rate) decreases with time t. Some of the main reasons for the occurrence of failures in this region are inadequate quality control, poor processes, substandard materials and workmanship, poor manufacturing methods, inadequate debugging, and human error.

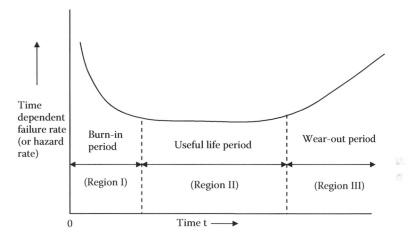

FIGURE 11.1
Bathtub hazard rate curve.

- During the useful life period (region II), the hazard rate remains constant with respect to time t. Some of the main reasons for the occurrence of failures in this region are higher random stress than expected, undetectable defects, human errors, low safety factors, abuse, and natural failures.

- Finally, during the wear-out period (region III), the hazard rate increases with time t. The main causes for the occurrence of failures in this region include poor maintenance, wear due to aging, wrong overhaul practices, short designed-in life of the item under consideration, wear due to friction and corrosion, and creep.

11.4 General Reliability, Mean Time to Failure, and Hazard Rate Formulas

A number of general formulas are commonly used to perform reliability analysis. Three of these formulas are presented next.

11.4.1 General Formula for Reliability

This general formula is expressed by [17]

$$R(t) = e^{-\int_0^t \lambda(t)\, dt} \tag{11.1}$$

where
   $R(t)$ is reliability at time t.
   t is time.
   $\lambda(t)$ is time-dependent failure rate (i.e., hazard rate).

Example 11.1
Assume that the hazard rate of an engineering system is given by

$$\lambda(t) = \lambda \qquad (11.2)$$

where λ is engineering system constant failure rate. Obtain an expression for the engineering system reliability by using Equation (11.1).
   Using Equation (11.2) in Equation (11.1) yields

$$R(t) = e^{-\int_0^t \lambda \, dt} \qquad (11.3)$$
$$= e^{-\lambda t}$$

Thus, Equation (11.3) is the expression for the engineering system reliability.

11.4.2 General Formula for Mean Time to Failure

This general formula can be expressed in the three different ways that follow [10]:

$$MTTF = \int_0^\infty R(t) \, dt \qquad (11.4)$$

or

$$MTTF = \lim_{s \to 0} R(s) \qquad (11.5)$$

or

$$MTTF = \int_0^\infty t \, f(t) \, dt \qquad (11.6)$$

where
   $f(t)$ is the failure or probability density function.
   s is the Laplace transform variable.
   $R(s)$ is the Laplace transform of $R(t)$.
   $MTTF$ is mean time to failure.

Example 11.2
Assume that the reliability of a piece of engineering equipment is expressed by

$$R(t) = e^{-\lambda t} \qquad (11.7)$$

where

t is time.

λ is engineering equipment failure rate.

Obtain an expression for the engineering equipment mean time to failure.
By substituting Equation (11.7) into Equation (11.4), we get

$$MTTF = \int_0^\infty e^{-\lambda t}\, dt$$

$$= \frac{1}{\lambda}$$

(11.8)

Thus, Equation (11.8) is the expression for the engineering equipment mean time to failure.

11.4.3 General Formula for Hazard Rate

This general formula can be expressed in the following three ways [10]:

$$\lambda(t) = \frac{f(t)}{1 - \int_0^t f(t)\, dt}$$

(11.9)

or

$$\lambda(t) = \frac{f(t)}{R(t)}$$

(11.10)

or

$$\lambda(t) = -\frac{1}{R(t)} \cdot \frac{dR(t)}{dt}$$

(11.11)

Example 11.3

Using Equation (11.7), obtain a hazard rate expression for the engineering equipment. Comment on the resulting expression.

Using Equation (11.7) in Equation (11.11) yields

$$\lambda(t) = -\frac{1}{e^{-\lambda t}} \frac{de^{-\lambda t}}{dt}$$

$$= \lambda$$

(11.12)

Thus, Equation (11.12) is the expression for the engineering equipment hazard rate. Note from this expression that the hazard rate is independent of time. Thus, it is simply referred to as the constant failure rate.

11.5 Common Reliability Networks

Engineering systems can form various types of configurations or networks in performing reliability analysis. Some of the commonly occurring of these networks are presented next.

11.5.1 Series Network

This is the simplest and probably the most commonly occurring reliability network in engineering systems. Its block diagram is shown in Figure 11.2. Each block in the figure denotes a unit or component. More specifically, the Figure 11.2 diagram represents a system composed of m units in series. If any one of the units fails, the system fails. In other words, all system units must work normally for the system to succeed.

If we let E_j denote the event that the jth unit in Figure 11.2 is successful, then the series system reliability is expressed by [5]

$$R_S = P(E_1 E_2 E_3 \ldots E_m) \tag{11.13}$$

where

R_S is the series system reliability.
$P(E_1 E_2 E_3 \ldots E_m)$ is probability of occurrence of events E_1, E_2, E_3,..., and E_m

For independent units, Equation (11.13) becomes

$$R_S = P(E_1)\, P(E_2)\, P(E_3) \ldots P(E_m) \tag{11.14}$$

where $P(E_j)$ is probability of occurrence of event E_j for $j = 1, 2, 3,\ldots, m$.
If we let $R_j = P(E_j)$ for $j = 1, 2, 3,\ldots, m$, Equation (11.14) becomes

$$R_S = R_1 R_2 R_3 \ldots R_m \tag{11.15}$$

where R_j is the unit j reliability for $j = 1, 2, 3,\ldots, m$.
For constant failure rate, λ_j, of unit j, using Equation (11.1), the reliability of the unit j is given by

$$R_j(t) = e^{-\int_0^t \lambda_j\, dt} \tag{11.16}$$

$$= e^{-\lambda_j t}$$

where $R_j(t)$ is reliability of unit j at time t.

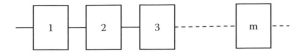

FIGURE 11.2
Block diagram of a series system containing m units.

Thus, by inserting Equation (11.16) into Equation (11.15), we obtain

$$R_S(t) = e^{-\sum_{j=1}^{m} \lambda_j t} \tag{11.17}$$

where $R_S(t)$ is series system reliability at time t.

By substituting Equation (11.17) into Equations (11.4) and (11.11), we get the following equations for the series system mean time to failure and hazard rate, respectively:

$$MTTF_S = \int_0^{\infty} e^{-\sum_{j=1}^{m} \lambda_j t} dt \tag{11.18}$$

$$= \frac{1}{\sum_{j=1}^{m} \lambda_j}$$

and

$$\lambda_S = -\frac{1}{e^{-\sum_{j=1}^{m} \lambda_j t}} \left[-\sum_{j=1}^{m} \lambda_j \right] e^{-\sum_{j=1}^{m} \lambda_j t} \tag{11.19}$$

$$= \sum_{j=1}^{m} \lambda_j$$

where

$MTTF_S$ is series system mean time to failure.

λ_S is series system hazard or failure rate.

Example 11.4

Assume that an engineering system is composed of three independent subsystems in series. The failure rates of subsystems 1, 2, and 3 are 0.005 failure/hour, 0.004 failure/hour, and 0.003 failure/hour, respectively. Calculate the following:

- engineering system reliability during a 40-hour mission;
- engineering system mean time to failure; and
- engineering system hazard rate.

By substituting the specified data values into Equations (11.17), (11.18), and (11.19), we get

$$R_S(40) = e^{-(0.005+0.004+0.003)(40)}$$

$$= 0.6188,$$

$$MTTF_S = \frac{1}{(0.005 + 0.004 + 0.003)}$$

$$= 83.33 \text{ hours},$$

and

$$\lambda_S = (0.005 + 0.004 + 0.003)$$
$$= 0.012 \text{ failures/hour}$$

Thus, the engineering system reliability, mean time to failure, and hazard rate are 0.6188, 83.33 hours, and 0.012 failures/hour, respectively.

11.5.2 Parallel Network

In this case, all units are active and at least one of these units must function normally for the system success. The block diagram of a parallel system containing m units is shown in Figure 11.3. Each block in the figure denotes a unit.

If we let $\bar{E}_j$ denote the event that the jth unit in Figure 11.3 is unsuccessful, then the parallel system failure probability is given by [5]

$$F_p = P(\bar{E}_1 \bar{E}_2 ... \bar{E}_m) \tag{11.20}$$

where

F_p is parallel system failure probability.
$P(\bar{E}_1 \bar{E}_2 ... \bar{E}_m)$ is probability of occurrence of failure events $\bar{E}_1, \bar{E}_2, ..., \bar{E}_m$.

For independent units, Equation (11.20) becomes

$$F_p = P(\bar{E}_1) P(\bar{E}_2) ... P(\bar{E}_m) \tag{11.21}$$

where $P(\bar{E}_j)$ is probability of occurrence of failure event $\bar{E}_j$; $j = 1, 2, ..., m$.

If we let $F_j = P(\bar{E}_j)$ for $j = 1, 2, ..., m$ in Equation (11.21) and then subtract the resulting equation from unity, we get the following expression for the parallel system reliability:

$$R_p = 1 - F_1 F_2 ... F_m \tag{11.22}$$

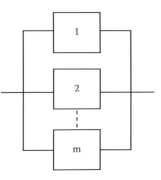

FIGURE 11.3
Block diagram of a parallel system containing m units.

where
R_p is parallel system reliability.
F_j is failure probability of unit j for $j = 1, 2,..., m$.

For constant failure rate, λ_j, of unit j, by subtracting Equation (11.16) from unity and then substituting it into Equation (11.22), we get

$$R_p(t) = 1 - \prod_{j=1}^{m} (1 - e^{-\lambda_j t})$$

(11.23)

where $R_p(t)$ is parallel system or network reliability at time t.

For identical units, by substituting Equation (11.23) into Equation (11.4), we get the following expression for the parallel system or network mean time to failure:

$$MTTF_p = \int_0^\infty [1 - (1 - e^{-\lambda t})^m] dt$$
$$= \frac{1}{\lambda} \sum_{j=1}^{m} \frac{1}{j}$$

(11.24)

where
$MTTF_p$ is mean time to failure of the parallel system with identical units.
λ is unit failure rate.

Example 11.5
An engineering system is composed of three independent, active, and identical units; at least one of the units must operate normally for system success. The unit failure rate is 0.0002 failure/hour. Calculate

- engineering system reliability for a 100-hour mission; and
- engineering system mean time to failure.

By substituting the given data values into Equations (11.23) and (11.24), we get

$$R_p(100) = 1 - [1 - e^{-(0.0002)(100)}][1 - e^{-(0.0002)(100)}][1 - e^{-(0.0002)(100)}]$$
$$= 0.9406$$

and

$$MTTF_p = \frac{1}{(0.0002)} \left[1 + \frac{1}{2} + \frac{1}{3} \right]$$
$$= 9{,}166.7 \text{ hours}$$

Thus, the engineering system reliability and mean time to failure are 0.9406 and 9,166.7 hours, respectively.

11.5.3 *K*-out-of-*m* Network

In this case, all m units are active and at least K units out of these m units must work normally for the system success. The parallel and series networks are the special cases of this network for $K = 1$ and $K = m$, respectively.

Using the binomial distribution for independent and identical units, the K-out-of m network or system reliability is expressed by [5]

$$R_{K/m} = \sum_{j=K}^{m} \binom{m}{j} R^j (1-R)^{m-j} \tag{11.25}$$

where

$$\binom{m}{j} = \frac{m!}{(m-j)!\, j!} \tag{11.26}$$

R is unit reliability.
$R_{K/m}$ is K-out-of-m network or system reliability.

For constant failure rate, λ, of each unit, by substituting Equation (11.3) into Equation (11.25), we get

$$R_{K/m}(t) = \sum_{j=K}^{m} \binom{m}{j} e^{-j\lambda t} (1-e^{-\lambda t})^{m-j} \tag{11.27}$$

where $R_{K/m}(t)$ is K-out-of-m network or system reliability at time t.
Using Equation (11.27) in Equation (11.4) yields

$$MTTF_{K/m} = \int_{0}^{\infty} \left[\sum_{j=K}^{m} \binom{m}{j} e^{-j\lambda t} (1-e^{-\lambda t})^{m-j} \right] dt$$

$$= \frac{1}{\lambda} \sum_{j=K}^{m} \frac{1}{j} \tag{11.28}$$

where $MTTF_{K/m}$ is K-out-of-m network or system mean time to failure.

Example 11.6

Assume that an engineering system is composed of four independent and identical units in parallel. At least two units must operate normally for the system's success. Calculate the engineering system mean time to failure if the failure rate of each unit is 0.0008 failure/hour.

By substituting the specified data values into Equation (11.28), we get

$$MTTF_{3/4} = \frac{1}{(0.0008)} \sum_{j=2}^{4} \frac{1}{j}$$

$$= \frac{1}{(0.0008)} \left[\frac{1}{2} + \frac{1}{3} + \frac{1}{4} \right]$$

$$= 1354.16 \text{ hours}$$

Thus, the engineering system mean time to failure is 1354.16 hours.

11.5.4 Standby System

This is another type of redundancy used to improve system reliability. In this case, the system has a total of $(m + 1)$ units, but only one unit operates. The remaining m units are kept in their standby mode. As soon as the operating unit fails, the switching mechanism detects the failure and turns on one of the m standby units.

The system fails when all the standby units fail. The system reliability for independent and identical units, time-dependent unit failure rate, and perfect switching mechanism and standby units is given by [5]

$$R_{ss}(t) = \sum_{j=0}^{m} \left[\left[\int_0^t \lambda(t)\,dt \right]^j e^{-\int_0^t \lambda(t)\,dt} \middle/ j! \right] \tag{11.29}$$

where
  $R_{ss}(t)$ is standby system reliability at time t.
  m is total number of standby units.
  $\lambda(t)$ is unit time-dependent failure rate or hazard rate.

For constant unit failure rate (i.e., $\lambda(t) = \lambda$), Equation (11.29) becomes

$$R_{ss}(t) = \sum_{j=0}^{m} (\lambda t)^j e^{-\lambda t} / j! \tag{11.30}$$

By inserting Equation (11.30) into Equation (11.4), we get

$$MTTF_{ss} = \int_0^{\infty} \left[\sum_{j=0}^{m} (\lambda t)^j e^{-\lambda t} / j! \right] dt$$

$$= \frac{m+1}{\lambda} \tag{11.31}$$

Example 11.7

A standby system has three independent and identical units: one operating and two on standby. The failure rate of each unit is 0.0002 failure/hour. Calculate the standby system mean time to failure if the standby units remain as good as new in their standby mode and the switching mechanism is perfect.

By substituting the given data values into Equation (11.31), we get

$$MTTF_{SS} = \frac{(2+1)}{(0.0002)}$$
$$= 15,000 \text{ hours}$$

Thus, the standby system mean time to failure is 15,000 hours.

11.6 Reliability and Maintainability Relationship

In order to have a clear understanding of the relationship between reliability and maintainability, some of the important aspects of both reliability and maintainability are discussed next.

11.6.1 Reliability

This is a design characteristic that results in durability of the system or item to perform its specified mission subject to stated conditions and time period. It is accomplished through actions such as choosing optimum engineering principles, satisfactory component sizing, controlling processes, and testing. The following are some specific general principles of reliability [5,17]:

- Design to minimize the occurrence of failures.
- Provide fail-safe designs.
- Design for simplicity.
- Provide redundancy when required.
- Use fewer numbers of parts to perform multiple functions.
- Minimize stress on parts.
- Provide for simple periodic adjustment of parts subject to wear.
- Maximize the use of standard parts.
- Use parts with proven reliability.
- Provide satisfactory safety factors between strength and peak stress values.

11.6.2 Maintainability

This is a built-in design and installation characteristic. It provides the end product an inherent ability to be maintained, thus ultimately leading to improved mission availability and reduction in maintenance cost, required tools and equipment, and required man-hours and skill levels. Some of the specific general principles of maintainability include [5,17]:

- Reduce life cycle maintenance costs.
- Reduce the amount, frequency, and complexity of required maintenance tasks.
- Reduce mean time to repair.
- Establish the extent of preventive maintenance to be performed.
- Reduce or eliminate altogether the need for maintenance.
- Reduce the amount of supply supports required.
- Consider benefits of modular replacement versus part repair or throwaway.
- Provide for maximum interchangeability.

11.7 Maintainability Measures

Various maintainability measures are used during the design phase to produce effective products with respect to maintainability. Some of these measures are mean time to repair (MTTR), the probability of completing repair in given time interval (i.e., the maintainability function); mean preventive maintenance time; and maximum corrective maintenance time. All these measures are presented next [5,7,17,18].

11.7.1 Mean Time to Repair (MTTR)

This is probably the most widely used maintainability measure or parameter in maintainability analysis. It is sometimes called mean corrective maintenance time. The system mean time to repair is defined by [5]

$$MTTR = \left[\sum_{i=1}^{m} \lambda_i t_i \right] \Big/ \sum_{i=1}^{m} \lambda_i \qquad (11.32)$$

where
  $MTTR$ is system mean time to repair.
  m is number of units.
  λ_i is constant failure rate of unit i for $i = 1, 2, 3,..., m$.
  t_i is time required to repair unit i for $i = 1, 2, 3,..., m$.

Example 11.8

Assume that an engineering system is composed of three nonidentical subsystems—1, 2, and 3—with constant failure rates $\lambda_1 = 0.0006$ failure/hour, $\lambda_2 = 0.0005$ failure/hour, and $\lambda_3 = 0.0004$ failure/hour, respectively. Corrective maintenance times of subsystems 1, 2, and 3 are 4, 3, and 2 hours, respectively. Calculate the engineering system mean time to repair.

By substituting the given data values into Equation (11.32), we get

$$MTTR = \frac{(0.0006)(4) + (0.0005)(3) + (0.0004)(2)}{(0.0006 + 0.0005 + 0.0004)}$$

$$= 3.133 \text{ hours}$$

Thus, the engineering system mean time to repair is 3.133 hours.

11.7.2 Maintainability Function

This is concerned with determining the probability of completing repair in a specified time interval. For a known repair time distribution, the maintainability function can be obtained by using the following equation [5,18]:

$$m(t) = \int_0^t f_r(t)\, dt \tag{11.33}$$

where

$m(t)$ is maintainability function (i.e., the probability that repair will be accomplished in time t when it starts at time $t = 0$).
t is time.
$f_r(t)$ is probability density function of the repair times.

Example 11.9

Assume that the repair times of a system are defined by the following probability density function (i.e., the repair times are exponentially distributed):

$$f_r(t) = \frac{1}{MTTR}\, e^{-\left[\frac{1}{MTTR}\right]t} \tag{11.34}$$

where

$f_r(t)$ is probability density function of the system repair times.
t is time.
$MTTR$ is system mean time to repair.

Obtain an expression for the maintainability function and calculate the probability that a repair will be accomplished in 3 hours if the system mean time to repair is 4 hours.

By substituting Equation (11.34) into Equation (11.33), we get

$$m(t) = \int_0^t \left[\frac{1}{MTTR} e^{-\left[\frac{1}{MTTR}\right]t} \right] dt$$

$$= 1 - e^{-\left[\frac{1}{MTTR}\right]t}$$

(11.35)

Using the given data values in Equation (11.35) yields

$$m(3) = 1 - e^{-\left(\frac{1}{4}\right)(3)}$$

$$= 0.5276$$

Thus, the expression for the maintainability function is given by Equation (11.35) and the probability that the system repair will be accomplished in 3 hours is 0.5276.

11.7.3 Mean Preventive Maintenance Time

This is a quite useful maintainability measure expressed by [5,18]

$$T_{mp} = \left[\sum_{i=1}^{K} t_{pi} f_{pi} \right] \Big/ \sum_{i=1}^{K} f_{pi}$$

(11.36)

where

T_{mp} is mean preventive maintenance time.
K is number of preventive maintenance tasks.
t_{pi} is elapsed time for preventive maintenance task i for $i = 1, 2, 3,..., K$.
f_{pi} is frequency of preventive maintenance task i for $i = 1, 2, 3,..., K$.

During the computation of T_{mp}, note that if the frequencies, f_{pi}, are specified in maintenance tasks per hour, then the values of t_{pi} must also be expressed in hours.

11.7.4 Maximum Corrective Maintenance Time

This maintainability measure is concerned with estimating the time to complete a specified percentage of all potential repair actions. Usually, the specified percentiles are the 90th and 95th. Because the estimation of maximum corrective maintenance time depends on the probability distribution describing the times to repair, equations for estimating maximum corrective maintenance time for three probability distributions are presented next [5,18].

11.7.4.1 Exponential

In this case, the maximum corrective maintenance time is expressed by

$$MT_{Cm} = \alpha (MTTR)$$

(11.37)

where
   MT_{Cm} is maximum corrective maintenance time.
   MTTR is mean time to repair.
   α is a constant whose values are 2.312 and 3 for the 90th and 95th percentiles, respectively.

11.7.4.2 Normal

In this case, the maximum corrective maintenance time is defined by

$$MT_{Cm} = MTTR + \theta \sigma_n \qquad (11.38)$$

where
   θ is a constant and its values are 1.28 and 1.65 for the 90th and 95th percentiles, respectively.
   σ_n is standard deviation of the repair times.

11.7.4.3 Lognormal

In this case, the maximum corrective maintenance time is expressed by

$$MT_{Cm} = anti \log (T_a + \theta \sigma_\ell) \qquad (11.39)$$

where
   T_a is mean of the logarithms of repair times.
   σ_ℓ is standard deviation of the logarithms of the repair times.

Additional information on the maximum corrective maintenance time is available in Dhillon [5,8].

11.8 System Availability and Unavailability

Availability and unavailability of a system are given by [5,18]

$$AV_S(t) = \frac{1}{(\lambda_S + \mu_S)} [\mu_S + \lambda_S e^{-(\lambda_S + \mu_S)t}] \qquad (11.40)$$

and

$$UAV_S(t) = \frac{\lambda_S}{\lambda_S + \mu_S} [1 - e^{-(\lambda_S + \mu_S)t}] \qquad (11.41)$$

where
$AV_S(t)$ is system availability at time t.
t is time.
λ_S is system constant failure rate.
μ_S is system constant repair rate.
$UAV_S(t)$ is system unavailability at time t.

As time t becomes very large, Equations (11.40) and (11.41) reduce to

$$AV_S = \lim_{t \to \infty} AV_S(t) = \frac{\mu_S}{\lambda_S + \mu_S} \qquad (11.42)$$

and

$$UAV_S = \lim_{t \to \infty} UAV_S(t) = \frac{\lambda_S}{\lambda_S + \mu_S} \qquad (11.43)$$

where
AV_S is system steady-state availability.
UAV_S is system steady-state unavailability.

Because

$$\lambda_S = \frac{1}{MTTF_S}$$

and

$$\mu_S = \frac{1}{MTTR_S}$$

Equations (11.42) and (11.43) become

$$AV_S = \frac{MTTF_S}{MTTF_S + MTTR_S} = \frac{System\ uptime}{System\ uptime + System\ downtime} \qquad (11.44)$$

and

$$UAV_S = \frac{MTTR_S}{MTTF_S + MTTR_S} = \frac{System\ downtime}{System\ uptime + System\ downtime} \qquad (11.45)$$

where
$MTTF_S$ is system mean time to failure.
$MTTR_S$ is system mean time to repair.

Example 11.10

An engineering system mean time to failure and mean time to repair are 400 hours and 20 hours, respectively. Calculate the system steady-state unavailability.

By substituting the given data values into Equation (11.45), we get

$$UAV_S = \frac{20}{400 + 20} = 0.0476$$

Thus, the engineering system unavailability is 0.0476.

11.9 Reliability and Maintainability Tools

Many methods are used to perform various types of reliability and maintainability analyses. Three of these methods that can be used to perform both reliability and maintainability analyses are as follows:

- failure modes and effect analysis (FMEA);
- fault tree analysis; and
- cause and effect diagram.

Each of these methods is described next.

11.9.1 Failure Modes and Effect Analysis (FMEA)

This is an important tool to evaluate engineering design at the initial stage from the reliability and maintainability aspects. FMEA was developed in the early 1950s for evaluating the design of flight control systems [19].

It helps to identify the need for and effects of design change and demands listing of potential failure modes of all system or equipment components on paper and their effects on the listed subsystems. The main steps in performing FMEA are shown in Figure 11.4. FMEA is called failure modes, effects, and criticality analysis (FMECA) when criticalities are assigned to failure mode effects. Additional information on FMEA is available in Dhillon [5].

11.9.2 Fault Tree Analysis

This is one of the most widely used methods for performing system reliability analysis; it arranges fault events in a tree-shaped diagram (thus, the name). The method is well suited to determine the combined effects of multiple failures. It was originally developed to evaluate the reliability of the Minuteman launch control system at Bell Telephone Laboratories in the early 1960s [5,20].

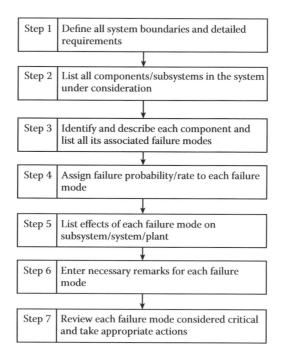

Step 1	Define all system boundaries and detailed requirements
Step 2	List all components/subsystems in the system under consideration
Step 3	Identify and describe each component and list all its associated failure modes
Step 4	Assign failure probability/rate to each failure mode
Step 5	List effects of each failure mode on subsystem/system/plant
Step 6	Enter necessary remarks for each failure mode
Step 7	Review each failure mode considered critical and take appropriate actions

FIGURE 11.4
Steps for performing FMEA.

The fault tree analysis starts by identifying an undesirable event—called the "top event"—associated with a system under consideration. The fault events that can cause the occurrence of the top event are generated and connected by logic gates known as OR, AND, etc. The construction of a fault tree of a system proceeds by generation of fault events (by asking, "How can this event occur?") successively until the fault events need not be developed further. These events are called elementary or primary events.

Overall, a fault tree may simply be described as the logic structure relating the primary fault events to the top event. Additional information on fault tree analysis is available in Dhillon [5] and Dhillon and Singh [21].

11.9.3 Cause and Effect Diagram

This is a quite useful approach for determining the root cause of a given problem and generating relevant ideas. Other names used for this approach are Ishikawa diagram, after its Japanese originator K. Ishikawa, and "fish bone" diagram because of its resemblance to the skeleton of a fish (as shown in Figure 11.5).

It can be seen from this figure that the right side (i.e., the fish head or the box) represents the effect (the problem or goal) and the dotted box on the left side contains "fish bones" that can be any set of factors considered to be important causes.

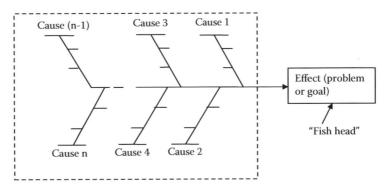

FIGURE 11.5
Cause and effect diagram layout.

The following basic steps are used to develop a cause and effect diagram:

- Develop problem statement.
- Brainstorm to identify all possible causes.
- Develop important cause classifications by stratifying into natural groupings and process steps.
- Develop the diagram.
- Refine all cause classifications by asking questions such as "What causes this?" and "Why does this condition exist?"

Additional information on the cause and effect diagram is available in Dhillon [5,18].

Problems

　　1. Discuss reliability and maintainability history.
　　2. Define the following terms:
　　　　- availability;
　　　　- reliability;
　　　　- maintainability; and
　　　　- useful life.
　　3. Describe the bathtub hazard rate curve.
　　4. Write three different general formulas for obtaining mean time to failure.
　　5. Obtain an expression for a parallel system hazard rate by using Equation (11.23).
　　6. List at least 10 general principles of reliability.

7. Describe two methods that can be used to perform reliability and maintainability analyses.
8. Prove that the sum of Equations (11.40) and (11.41) is equal to unity.
9. Assume that an engineering system is composed of four nonidentical subsystems—1, 2, 3, and 4—with constant failure rates $\lambda_1 = 0.0001$ failure/hour, $\lambda_2 = 0.0002$ failure/hour, $\lambda_3 = 0.0003$ failure/hour, and $\lambda_4 = 0.0004$ failure/hour, respectively. Calculate the engineering system mean time to repair if the corrective maintenance times of subsystems 1, 2, 3, and 4 are 2, 4, 6, and 8 hours, respectively.
10. Assume that an engineering system is composed of five independent and identical units in parallel. At least three units must operate normally for the system success. Calculate the engineering system mean time to failure if the constant failure rate of each unit is 0.004 failure/hour.

References

1. Lyman, W. J. 1933. Fundamental consideration in preparing a master system plan. *Electrical World* 101:778–792.
2. Smith, S. A. 1934. Service reliability measured by probabilities of outage. *Electrical World* 103:371–374.
3. Dhillon, B. S. 1983. *Power system reliability, safety, and management.* Ann Arbor, MI: Ann Arbor Science Publishers.
4. Coppola, A. 1984. Reliability engineering of electronic equipment: A historical perspective. *IEEE Transactions on Reliability* 33:29–35.
5. Dhillon, B. S. 1999. *Design reliability: Fundamentals and applications.* Boca Raton, FL: CRC Press.
6. Dhillon, B. S. 2007. *Applied reliability and quality: Fundamentals, methods, and procedures.* London: Springer–Verlag.
7. AMCP 706-133. 1976. *Engineering design handbook: Maintainability engineering theory and practice.* Washington, D.C.: Department of Defense.
8. Moss, M. A. 1985. *Minimal maintenance expense.* New York: Marcel Dekker, Inc.
9. Retterer, B. L., and Kowalski, R. A. 1984. Maintainability: A historical perspective. *IEEE Transactions on Reliability* 33:56–61.
10. Shooman, M. L. 1968. *Probabilistic reliability: An engineering approach.* New York: McGraw–Hill Book Company.
11. Dhillon, B. S. 1993. *Reliability and quality control: Bibliography on general and specialized areas.* Gloucester, Ontario, Canada: Beta Publishers, Inc.
12. Dhillon, B. S. 1993. *Reliability engineering applications: Bibliography on important application areas.* Gloucester, Ontario, Canada: Beta Publishers, Inc.
13. MIL-STD-721. 1974. Definitions of effectiveness terms for reliability, maintainability, human factors, and safety. Washington, D.C.: Department of Defense.

14. Naresky, J. J. 1970. Reliability definitions. *IEEE Transactions on Reliability* 19:198–200.
15. Omdahl, T. P., ed. 1988. *Reliability, availability, and maintainability (RAM) dictionary.* Milwaukee, WI: ASQC Quality Press.
16. Von Alven, W. H., ed. 1964. *Reliability engineering.* Englewood Cliffs, NJ: Prentice Hall, Inc.
17. AMCP-706-134. 1972. Maintainability guide for design. Prepared by the Department of the Army, Department of Defense, Washington, D.C.
18. Dhillon, B. S. 1999. *Engineering maintainability: How to design for reliability and easy maintenance.* Houston, TX: Gulf Publishing Company.
19. Countinho, J. S. 1964. Failure effect analysis. *Transactions of the New York Academy of Sciences* 26:564–584.
20. Haasl, D. F. 1965. Advanced concepts in fault tree analysis. *Proceedings of the System Safety Symposium.* Available from the University of Washington Library, Seattle, WA.
21. Dhillon, B. S., and Singh, C. 1981. *Engineering reliability: New techniques and applications.* New York: John Wiley & Sons.

Bibliography: Literature on Life Cycle Costing

Introduction

Over the years, a large number of publications on various aspects of life cycle costing have appeared in the form of journal articles, conference proceedings articles, books, etc. This bibliography presents an extensive list of such publications. The period covered by the listing is from 1988 to 2008. The main objective of this listing is to provide readers with sources for obtaining additional information on life cycle costing.

Publications

Abraham, D. M. 2003. Life cycle cost integration for the rehabilitation of wastewater infrastructure. *Proceedings of the Construction Research Congress* 627–635.

Adler, D., Willman, T., and Lilly, E. 1997. Figuring life-cycle costs in the real world. *Chemical Processing* 60 (8): 29–32.

Adler, D. J., Herkamp, J. A., Wiesler, J. R., and Williams, S. B. 1995. Life cycle cost and benefits of process automation in bulk pharmaceuticals. *ISA Transactions* 34 (2): 133–139.

Ahmed, N. U. 1995. A design and implementation model for life cycle cost management system. *Information and Management* 28, (4): 261–269.

Akselsson, H., and Burstrom, B. 1994. Life cycle cost procurement of Swedish State Railways' high-speed train X2000. Proceedings of the Institution of Mechanical Engineers, Part F: *Journal of Rail and Rapid Transit* 208 (1): 51–59.

Aktacir, M. et al. 2006. Life-cycle cost analysis for constant-air-volume and variable-air-volume air-conditioning systems. *Applied Energy* 83 (6): 606–627.

Alfredsson, K. 2001. Life cycle cost in focus. *Water and Wastewater International* 16 (2): 25.

Ali Khan Malik, M., and Kolodchak, P. 1990. Cost-reliability relationship in life cycles. *Proceedings of the International Industrial Engineering Conference* 581–586.

Allen, E. C. et al. 2006. Mission-based simulation software development for optimizing air vehicle life cycle costs. *Proceedings of the AIAA Modeling and Simulation Technologies Conference* 1021–1031.

Anon. 1993. Effect of life cycle costs versus initial costs. *NASA Reference Publication* 1310:146.

———. 1994. Life cycle costing. *Steel Times* 222 (1): 21–22.

———. 1995. Improving gas turbine availability and life cycle costs. *International Power Generation* 18 (5): 38–39, 42.

————. 1995. Life cycle costing: Report on the introduction of life cycle costing techniques to the selection of maintenance coating systems for offshore fabric. *Offshore Engineer* 7:30.

————. 1995. Reader responds to life cycle costing article. *Chemical Processing* 58 (8): 12.

————. 1995. Thermal insulation environmental impacts and life-cycle costs. *Construction Specifier* 48 (6): 64–69.

————. 1996. Do life-cycle costs validate the standard-plant concept? *Power* 140 (8): 126.

————. 1996. Life-cycle costing reveals masonry's long-term value. *Aberdeen's Magazine of Masonry Construction* 9 (12): 555.

————. 1996. Up time: New low-maintenance components from Eaton help reduce operating and life cycle costs. *Diesel Equipment Superintendent* 74 (4): 58.

————. 1997. GTX100 promises high reliability and low life cycle costs. *Modern Power Systems* 17 (7): 23, 25.

————. 1997. Life cycle cost implications of roofing decisions. *Interface* 15 (2): 7.

————. 1997. Life cycle costing proves concrete's economy. *Better Roads* January: 21–24.

————. 1997. Life-cycle costing provides economy. *Better Roads* 67 (1): 21.

————. 1997. Life cycle costs. *Aerospace Engineering* 17 (10): 29.

————. 1998. Bridge plans receive life-cycle costs. *ENR (Engineering News-Record)* 240 (20): 19.

————. 1998. It's time to calibrate financial models with real life-cycle costs. *Power* 142 (4): 4.

————. 1998. Life cycle cost analysis for pumping systems. *World Pumps* 383:28–32.

————. 1999. Procedures for welding titanium piping helping U.S. Navy to reduce ship life-cycle costs. *Welding Journal* 78 (4): 92.

————. 1999. State DOTs update life-cycle cost analysis. *Better Roads* 69 (10): 25.

————. 2000. Intelligent wells: Forecasting life-cycle costs. *Hart's E and P* 73 (8): 125.

————. 2000. ITT: Technology leadership and customer satisfaction driving life cycle cost. *World Pumps* April: 18–21.

————. 2000. Managing equipment life-cycle costs. *Chemical Engineering* 107 (2): 80.

————. 2000. Pump users' forum with a focus on life cycle costs. *World Pumps* 407:44.

————. 2001. Life-cycle strategy for pumps improves cost structure. *World Pumps* 413:30–32.

————. 2001. Roofing and life-cycle cost. *Buildings* 95 (5): 74.

————. 2002. A local authority thinks hard about life-cycle costs. *Highways* 71 (5): 32.

Arditi, D. A. et al. 1996. Life-cycle costing in municipal construction projects. *Journal of Infrastructure Systems* 2 (1): 5–14.

Arditi, D., and Messiha, H. M. 1999. Life cycle cost analysis (LCCA) in municipal organizations. *Journal of Infrastructure Systems* 5 (1): 1–10.

Arnold, B. D. et al. 2005. Life-cycle costing of air filtration. *ASHRAE Journal* 47 (11): 30–32.

Arpke, A., and Hutzler, N. 2005. Operational life-cycle assessment and life-cycle cost analysis for water use in multi-occupant buildings. *Journal of Architectural Engineering* 11 (3): 99–109.

Ashworth, A. 1989. Life-cycle costing: A practice tool? *Cost Engineering* 31 (3): 8–11.

Asiedu, Y., and Gu, P. 1998. Product life cycle cost analysis: State of the art review. *International Journal of Production Research* 36 (4): 883–908.

Balda, D. M., and Gustafson, D. A. 1990. Cost estimation models for the reuse and prototype software development life-cycles. *ACM SIGSOFT Software Engineering Notes* 15 (3): 42–50.

Baliwangi, L. et al. 2006. Optimizing ship machinery maintenance scheduling through risk analysis and life cycle cost analysis. *Proceedings of the 25th International Conference on Offshore Mechanics and Arctic Engineering* 127–133.

Bang, K. L. et al. 1996. Development of guidelines based on life-cycle cost to replace level-of-service concept in capacity analysis. *Transportation Research Record* 1572:9–17.

Barrick, M. D. 1989. Productivity and life cycle cost issues in applications of embedded fiber optic sensors in smart skins. *Proceedings of the SPIE Conference* 171–179.

Barros, L. L. 1998. The optimization of repair decisions using life cycle cost parameters. *IMA Journal of Mathematics Applied in Business and Industry* 9 (4): 403–413.

Battlebury, D. R. 1991. The practical application of life cycle costing to the design of power systems. *Proceedings of the Third International Conference on Probabilistic Methods Applied to Electric Power Systems* 6–8.

Bears, J., and Coathup, L. 1991. Evaluation of the life cycle cost for universal fiber access. *Proceedings of the First International Workshop on Photonic Networks, Components and Applications* 81–89.

Becker, E. A. et al. 1999. Life cycle cost of urban pavements. *Concrete Engineering International* 3 (3): 26–28.

Becker, S. 1998. Bringing advanced bogie technology to Europe will cut life-cycle costs. *Railway Gazette International* 154 (9): 599–600.

Bell, J. H. 1990. Parts recovery life cycle costs. *Proceedings of the Test Engineering Conference* 181–185.

Bell, P. I., and Trigger, J. P. 1998. Access network life-cycle costs. *BT Technology Journal* 16 (4): 165–174.

Bentz, E. J., Bentz, C. B., and O'Hora, T. D. 2001. Comparative assessment of low-level radioactive waste life-cycle disposal costs of U.S. commercial facilities. *Proceedings of the 8th International Conference on Radioactive Waste Management and Environmental Remediation* 751–757.

Bescherer, F. 2006. Towards the optimum cost of ownership of switched-mode power supplies: Early stage cost management with life-cycle costing. *Proceedings of the IEEE 32nd Annual Conference on Industrial Electronics* 2203–2207.

Bettigole, N. H. 1993. Bridge engineering and life cycle cost. *Proceedings of the Structures Congress* 1047–1052.

———. 1995. Bridge management and life cycle cost. *Proceedings of the Structures Congress* 668–669.

Bhaskaran, R., Palaniswamy, N., and Rengaswamy, N. S. 2006. Life-cycle cost analysis of a concrete road bridge across open sea. *Materials Performance* 45 (10): 51–55.

Birkenshaw, J. 2003. Life cycle costing of print on-demand digital printing of books and packaging materials. *Proceedings of the International Conference on Digital Production Printing and Industrial Applications* 12–13.

Blanchard, B. S. 1988. The measures of a system—performance, life-cycle cost, system effectiveness, or what? *Proceedings of the IEEE National Aerospace and Electronics Conference* 1434–1439.

Bodsberg, L., and Hokstad, P. 1995. A system approach to reliability and life-cycle cost of process safety systems. *IEEE Transactions on Reliability* 44 (2): 179–186.

Boehm, B. et al. 2004. A software product line life cycle cost estimation model. *Proceedings of the International Symposium on Empirical Software Engineering* 156–164.

Bohoris, G. A. 1993. Life-cycle costs and comparative statistical techniques for censored reliability data. *Journal of the Operational Research Society* 44 (4): 355–360.

Bonner, J. A. et al. 1989. Fossil power plant life cycle management as a least cost planning approach available to utility and industrial plant operators. *Proceedings of the American Power Conference* 956–958.

Botelho, D. et al. 2000. Life-cycle-cost-based design criteria for Gulf of Mexico minimum structures. *Proceedings of the Annual Offshore Technology Conference* 87–94.

Boussabaine, H. A., and Kirkham, R. J. 2004. *Whole life cycle costing: Risk and risk responses.* Oxford, England: Blackwell Publishing.

Breidenbach, D. P. 1989. Life cycle cost analysis. *Proceedings of the IEEE 1989 National Aerospace and Electronics Conference* 1216–1220.

Breniere, X., and Tribolet, P. 2006. IR detectors life cycle cost and reliability optimization for tactical applications. *Proceedings of the International Society for Optical Engineering Conference on Electro-Optical and Infrared Systems: Technology and Applications III* 63950D1–63950D12.

Brentlinger, L. A., Hofmann, P. L., Peterson, R. W., and Dippold, D. G. 1988. Transportable storage casks: An analysis of life cycle dose and life cycle cost. *Proceedings of the Summer Computer Simulation Conference* 742–746.

Brooks, S. M. 1996. Life cycle costs estimates for conceptual ideas. *Proceedings of the IEEE National Aerospace and Electronics Conference* 541–546.

Brown, D. R., and Humphreys, K. K. 1988. Battery life-cycle cost analysis. *Electric Vehicle Developments* 7 (3): 81–82.

Bruhwiler, E., and Adey, B. 2005. Improving the consideration of life-cycle costs in bridge decision-making in Switzerland. *Structure and Infrastructure Engineering* 1 (2): 145–157.

Bruzzone, A. G., Briano, C., Massei, M., and Poggi, S. 2006. Simulation and optimization as decision support system in relation to life cycle cost of new aircraft carriers. *Proceedings of the Sixth IASTED International Conference on Modeling, Simulation, and Optimization* 133–138.

Buncher, M., and Rosenberger, C. 2006. Understanding the true economics of using polymer modified asphalt through life cycle cost analysis. *Paving the Way* 8 (2): 1–20.

Burley, E., and Rigden, S. R. 1997. Use of life cycle costing in assessing alternative bridge design. *Proceedings of the Institution of Civil Engineers* 121 (1): 22–27.

Burns, D. J., and Formaniak, A. 1992. Reliability and life cycle cost evaluation for system design. *Proceedings of the Safety and Reliability Conference* 353–367.

Burrows, C. 2003. Designed for life—First cost or life. *Proceedings of the International Conference on New Trains* 19–26.

Burstrom, B., Ericsson, G., and Kjellsson, U. 1994. Verification of life-cycle cost and reliability for the Swedish high speed train X2000. *Proceedings of the Annual Reliability and Maintainability Symposium* 166–171.

Cain, J. P., Habash, N., and Gibson, J. A. 1994. Analysis of military systems using an interactive life cycle costing model. *Proceedings of the IEEE National Aerospace and Electronics Conference* 1218–1224.

Calvo, A. B., Danish, A. J., and Marcus, D. 2002. Web-LCCA: A life-cycle cost model for evaluation of COTS and custom display designs. *Proceedings of SPIE Conference* 70–80.

Cardullo, M. W. 1993.Total life-cycle cost analysis of conventional and alternative fueled vehicles. *IEEE Aerospace and Electronic Systems Magazine* 8 (11): 39–43.

———. 1995. Total life cycle cost model for electric power stations. *Proceedings of the 30th Intersociety Energy Conversion Engineering Conference* 409–414.

Carnahan, J. V., and Marsh, C. 1998. Comparative life-cycle cost analysis of underground heat distribution systems. *Journal of Transportation Engineering* 124 (6): 594–605.

Carriere, M., Schoenau, G. J., and Besant, R. W. 1998. Revised procedure for duct design with minimum life-cycle cost. *ASHRAE Transactions* 104 (2): 62–67.

Carrubba, E. R. 1992. Integrating life-cycle cost and cost of ownership in the commercial sector. *Proceedings of the Annual Reliability and Maintainability Symposium* 101–108.

Carter, M. F. 1998. Designing machine tools to minimize life cycle cost. ASME Design Engineering Division (publication) DE-99. *Applications of Design for Manufacturing* 1–5.

Cataldo, R. L., and Sefcik, R. J. 1993. Life cycle cost-A consideration for selection of advanced power systems. *Proceedings of the 28th Intersociety Energy Conversion Engineering Conference* 457–462.

Chafee, S. S. 1996. Fundamental requirements of life cycle costing: Projecting life cycle costs for electronic system modernization. *Proceedings of the IEEE Technical Applications Conference* 41–46.

Chang, S. E., Shinozuka, M., and Ballantyne, D. B. 1997. Life cycle cost analysis with natural hazard risk: A framework and issues for water systems. *Proceedings of the International Workshop on Optimal Performance of Civil Infrastructure Systems* 58–73.

Chang, S. E., and Shinozuka, M. 1996. Life-cycle cost analysis with natural hazard risk. *Journal of Infrastructure Systems* 2 (3): 118–126.

Cheng, F. Y. et al. 1999. Genetic algorithm for multi-objective optimization and life-cycle cost. *Proceedings of the Structures Congress* 484–489.

Chewning, I. M., and Moretto, S. J. 2000. Advances in aircraft carrier life cycle cost analysis for acquisition and ownership decision-making. *Naval Engineers Journal* 112 (3): 97–110.

Choi, J., and Bahia, H. U. 2004. Life-cycle cost analysis-embedded Monte Carlo approach for modeling pay adjustment at state departments of transportation. *Transportation Research Record* 1900:86–93.

Christensen, P. N., Sparks, G. A., and Kostuk, K. J. 2005. A method-based survey of life cycle costing literature pertinent to infrastructure design and renewal. *Canadian Journal of Civil Engineering* 32 (1): 250–259.

Christian, J. et al. 1998. Life cycle costs of the barrack block: Are we building better and smarter? *Proceedings of the Annual Conference of the Canadian Society for Civil Engineering* 129–138.

Christian, J., and Pandeya, A. N. 1995. Knowledge acquisition for life-cycle costs. *Proceedings of the Conference on Applications of Artificial Intelligence in Engineering* 273–280.

Clark, J. P. 1996. Life cycle analysis methodology incorporating private and social costs. *VDI Berichte* 1307:1–19.

Coathup, L., Goddard, G. W., McEachern, J., and Bears, J. 1990. Evaluation of the life cycle cost for universal fiber access. *Proceedings of the IEEE International Conference on Communications* 1100–1104.

Cole, P. A., Jr. 1991. The impact of manpower, personnel, and training (MPT) on life cycle cost. *Proceedings of the IEEE National Aerospace and Electronics Conference* 842–848.

Corotis, R. B., Ellis, J. H., and Jiang, M. X. 2005. Modeling of risk-based inspection, maintenance and life-cycle cost with partially observable Markov decision processes. *Structure and Infrastructure Engineering* 1 (1): 75–84.

Cosiol, J. 2001. Weighing the first and life-cycle costs of building control systems. *HPAC Heating, Piping, Air Conditioning Engineering* 73 (10): 9.

Cranford, E. L., III et al. 2002. Reduced life cycle costs and improved analysis accuracy utilizing Westem's integrated modeling methods. *ASME Pressure Vessels and Piping Division Publication (PVP)* 443 (1): 9–15.

Crawley, M. F., and Bell, J. M. 1993. Application of life-cycle cost analysis to pneumatic conveying systems. *Powder Handling & Processing* 5 (3): 213–218.

Crouch, V. 1994. High demand telemetry system that maximizes future expansion at minimum life-cycle cost. *Proceedings of the International Telemetering Conference* 26–33.

Curry, E. E. 1989. STEP: A tool for estimating avionics life cycle costs. *IEEE Aerospace and Electronics Systems Magazine* 4 (1): 30–32.

————. 1993. FALCCM-H: Functional avionics life cycle cost model for hardware. *Proceedings of the IEEE National Aerospace and Electronics Conference* 950–953.

Dacko, L. M., and Darlington, R. F. 1988. Life-cycle cost procedure for commercial aircraft subsystem. *Proceedings of the Annual Reliability and Maintainability Symposium* 389–394.

Dahlen, P., and Bolmsjo, G. S. 1996. Life-cycle cost analysis of the labor factor. *International Journal of Production Economics* 46–47, 459–467.

Dartnall, J., Adhikari, A. K., and McNab, J. 2006. Designing functional products in the best interest of the user—With a factor 10 reduction in life cycle cost—Example: A (solar) air conditioning system. *Proceedings of the ASME International Mechanical Engineering Congress and Exposition* 197–205.

Davis, N., Jones, J., and Warrington, L. 2003. A framework for documenting and analyzing life-cycle costs using a simple network based representation. *Proceedings of the Annual Reliability and Maintainability Symposium* 232–236.

De Boer, G., and Greidanus, D. 2006. Utilization of customized hydraulics to elongate pump life and lower life cycle costs. *Proceedings of the Institution of Mechanical Engineers Ninth European Fluid Machinery Congress on Applying the Latest Technology to New and Existing Process Equipment* 95–102.

De Haas, E. 1991. Reduced life-cycle cost through RMSH. *Proceedings of the 1991 IEEE/ASME Joint Railroad Conference* 23–25.

DellaVilla, S. A. et al. 2006. Parts life management—Essential for minimizing life cycle costs. *Proceedings of ASME Power Conference* 5.

Del Re, V., Lezzerini, L., Menna, E., Moro, F., Auer, C., and Bevilacqua, S. 2005. Neptune: A tool and an approach for life cycle cost reduction in space ground segment. *Proceedings of the 6th International Symposium on Reducing the Costs of Spacecraft Ground systems and Operations* 377–383.

DeLuchi, M., Wang, Q., and Sperling, D. 1989. Electric vehicles: Performance, life-cycle costs, emissions, and recharging requirements. *Transportation Research* 23A (3): 255–278.

Desai, A. R. 1997. Life-cycle cost estimating helps make turbine decisions. *Pipe Line & Gas Industry* 80 (10): 65–68.

Deschaine, L. M., Ades, M. J., Ahfeld, D. P., and O'Brien, D. 1998. An optimization algorithm to minimize the life cycle cost of implementing an aquifer remediation project-theory and case example. *Proceedings of the Simulators International Conference* 53–58.

De Vasconcellos, N. M., and Yoshimura, M. 1999. Life cycle cost model for acquisition of automated systems. *International Journal of Production Research* 37 (9): 2059–2076.

Devereux, B., and Singh, R. 1994. Use of computer simulation techniques to assess thrust rating as a means of reducing turbo-jet life cycle costs. *Proceedings of the International Gas Turbine and Aeroengine Congress* 1–8.

Dhillon, B. S. 1989. *Life cycle cost: Techniques, models, and applications.* New York: Gordon and Breach Science Publishers.

Dieffenbach, J. R., and Mascarin, A. E. 1993. Body-in-white material systems: A life-cycle cost comparison. *JOM* 45 (6): 16–19.

Dietrich, J. M. 2004. Life cycle process management for environmentally sound and cost effective semiconductor manufacturing. *Proceedings of the IEEE International Symposium on Electronics and the Environment* 168–172.

Di Martino, P., Rosi, R., and Zanetta, L. 1996. Life cycle costs comparison and sensitivity analysis for multimedia networks. *Proceedings of the European Conference on Networks and Optical Communications* 306–310.

Di Stefano, P. 2006. Tolerances analysis and cost evaluation for product life cycle. International *Journal of Production Research* 44 (10): 1943–1961.

Doswell, B. E. 1988. Who is doing what about life cycle costing. *Proceedings of the Conference on Military Computers, Graphics and Software* 333–344.

Dowdell, D. C. et al. 2000. An integrated life cycle assessment and cost analysis of the implications of implementing the proposed waste from electrical and electronic equipment (WEEE) directive. *Proceedings of the IEEE International Symposium on Electronics and the Environment* 1–10.

Dowlatshahi, S. 1997. Elements of time-based competition and life cycle costing in concurrent engineering environments. *Proceedings of the Annual Meeting of the Decision Sciences Institute* 1091–1092.

Durairaj, S. K. et al. 2002. Evaluation of life cycle cost analysis methodologies. *Corporate Environmental Strategy* 9 (1): 30–39.

Egan, W. F., and Iacovelli, J. W. 1996. Projected life cycle costs of an exterior insulation and finish system. *ASTM Special Technical Publication* 1269:189–207.

Ehlen, M. A. 1997. Life-cycle costs of new construction materials. *Journal of Infrastructure Systems* 3 (4): 129–133.

———. 1999. Life-cycle costs of fiber-reinforced-polymer bridge decks. *Journal of Materials in Civil Engineering* 11 (3): 224–230.

El Hayek, M., Van Voorthuysen, E., and Kelly, D. W. 2005. Optimizing life cycle cost of complex machinery with rotable modules using simulation. *Journal of Quality in Maintenance Engineering* 11 (4): 333–347.

El-Diraby, T. E. 2006. Web-services environment for collaborative management of product life-cycle costs. *Journal of Construction Engineering and Management* 132 (3): 300–313.

El-Diraby, T. E., and Rasic, I. 2004. Framework for managing life-cycle cost of smart infrastructure systems. *Journal of Computing in Civil Engineering* 18 (2): 115–119.

Embacher, R. A., and Snyder, M. B. 2001. Life-cycle cost comparison of asphalt and concrete pavements on low-volume roads: Case study comparisons. *Transportation Research Record* 1749:28–37.

Emblemsvag, J. 2003. *Life cycle costing: Using activity-based costing and Monte Carlo methods to manage future costs and risks.* New York: John Wiley & Sons.

Emblemsvag, J., and Bras, B. 1997. Method for life-cycle design cost assessments using activity-based costing and uncertainty. *Engineering Design and Automation* 3 (4): 339–354.

Erto, P., and Lanzotti, A. 1994. Statistical model "life cycle cost-reliability" for a new mass transit vehicle. *Proceedings of the International Conference on Computer Aided Design, Manufacture and Operation in the Railway and Other Mass Transit Systems* 101–110.

Esteves, J. M. et al. 2001. Towards an ERP life-cycle cost model. *Proceedings of the Information Resources Management Association International Conference* 431–435.

Fabrycky, W. J., and Blanchard, B. S. 1991. *Life cycle cost and economic analysis.* Englewood Cliffs, NJ: Prentice Hall.

Fagen, M. E., and Phares, B. M. 2000. Life-cycle cost analysis of a low-volume road bridge alternative. *Transportation Research Record* 1696:8–13.

Fairclough, M. R. 1989. ARENA—A software aid for assessing system availability and life cycle cost. *Proceedings of the Reliability '89 Conference* 4B/1/1–4B/1/6.

Fazio, V., Savio, S., and Firpo, P. 2001. EXCEL® based simulation procedure for complex systems life cycle cost estimation. *Proceedings of the European 15th Modeling and Simulation Conference* 48–53.

Fenton, S. 2000. LIFT: A vehicle for low life cycle costs. *ABB Review* 3:65–68.

Ferreira, A. 2005. A life-cycle cost analysis system for transportation asset management systems. *Proceedings of the 16th IASTED International Conference on Modeling and Simulation* 234–239.

Feuerherd, K. H. et al. 2001. Eco efficiency and target costing for making eco-design more effective: Integrating life cycle assessment and life cycle cost management. *Proceedings of the Second International Symposium on Environmentally Conscious Design and Inverse Manufacturing* 745–759.

Fiksel, J., and Wapman, K. 1994. How to design for environment and minimize life cycle cost. *Proceedings of the IEEE International Symposium on Electronics & the Environment* 75–80.

Finch, E. F. 1994. Uncertain role of life cycle costing in the renewable energy debate. *Renewable Energy* 5 (5–8): 1436–1443.

Fisher, G. B., Grunter, W., and Coudray, B. 1997. Consideration of reliability and life-cycle costs for future airborne radars. *Proceedings of the IEE Conference on Radar* 449:348–351.

Fitzpatrick, M., and Paasch, R. 1999. Analytical method for the prediction of reliability and maintainability based life-cycle labor costs. *Journal of Mechanical Design, Transactions of the ASME* 121 (4): 606–613.

Fixson, K. 2004. Assessing product architecture costing: Product life cycles, allocation rules, and cost models. *Proceedings of the ASME Design Engineering Technical Conference* 857–868.

Flintsch, W., and Chen, C. 2004. Soft computing-based infrastructure life-cycle cost analysis tools. *Proceedings of the ASCE Information Technology Symposium* 1–15.

Foerstemann, M., and Staudacher, S. 2004. Optimizing the architecture of civil turbofan engines to improve life cycle costs/value added. *Proceedings of the ASME Turbo Conference on Aircraft Engine, Ceramics, Controls, Diagnostics, and Instrumentation* 89–96.

Fragiadakis, M., Lagaros, N. D., and Papadrakakis, M. 2006. Performance-based multiobjective optimum design of steel structures considering life-cycle cost. *Structural and Multidisciplinary Optimization* 32 (1): 1–11.

Frangopol, D. M., and Furuta, H., eds. 2001. *Life cycle cost analysis and design of civil infrastructure systems.* Reston, VA: Structural Engineering Institute of the American Society of Civil Engineers.

Frangopol, D. M., and Lin, K. 1997. Reliability-based optimum design for minimum life-cycle cost. *Proceedings of the U.S.–Japan Joint Seminar on Structural Optimization* 67–78.

Frangopol, D. M., Lin, K. Y., and Estes, A. C. 1997. Life-cycle cost design of deteriorating structures. *Journal of Structural Engineering* 123 (10): 1390–1401.

Frangopol, D. M., and Liu, M. 2004. Life-cycle cost analysis for highways bridges: Accomplishments and challenges. *Proceedings of the Structures Congress* 25–33.

———. 2007. Maintenance and management of civil infrastructure based on condition, safety, optimization, and life-cycle cost. *Structure and Infrastructure Engineering* 3 (1): 29–41.

Frangopol, D. M. et al. 1997. Optimal bridge management based on lifetime reliability and life-cycle cost. *Proceedings of the International Workshop on Optimal Performance of Civil Infrastructure Systems* 98–115.

———. 1999. Integration of NDE in life-cycle cost of highway bridges. *Proceedings of the Structures Congress* 825–828.

Frenkel, V. 2003. Consider life-cycle costs in designing or upgrading water pretreatment systems. *Power Engineering* 107 (5): 43–45.

Fukuda, M., Watanabe, I., Terada, N., Shimazoe, T., and Okutani, T. 2005. A study of railway signaling system design method through requirement analysis and integrated life cycle cost evaluation. *Transactions of the Institute of Electrical Engineers of Japan, Part D* 125D (7): 681–690.

Furukawa, N. et al. 2006. Development of new steels to reduce life cycle costs of steel bridges and build application experience. *SEAISI Quarterly (South East Asia Iron and Steel Institute)* 35 (3): 22–35.

Furuta, H., Koyama, K., Oi, M., and Sugimoto, H. 2005. Life-cycle cost evaluation of multiple bridges in road network considering seismic risk. *Proceedings of the 6th International Bridge Engineering Conference on Reliability, Security, and Sustainability in Bridge Engineering* 343–347.

Furuta, H. et al. 1999. Life-cycle cost design of deteriorating bridges using genetic algorithm. *Proceedings of the Structures Congress* 243–246.

———. 2003. Life-cycle cost analysis for infrastructure systems: Life-cycle cost vs. safety level vs. service life. *Proceedings of the Life-Cycle Performance of Deteriorating Structures: Assessment, Design and Management Conference* 19–25.

———. 2005. Effects of seismic risk on life-cycle cost analysis for bridge maintenance. *Proceedings of the 4th International Conference on Current and Future Trends in Bridge Design, Construction, and Maintenance* 22–33.

———. 2006. Life-cycle cost design using improved multi-objective genetic algorithm. *Proceedings of the 17th Analysis and Computation Specialty Conference* 23–36.

Fwa, T. F., and Sinha, K. C. 1991. Pavement performance and life-cycle cost analysis. *Journal of Transportation Engineering* 117 (1): 33–46.

Garcia, H. F. 1989. Life cycle costing: An application of total cost purchasing in both public and private sectors. *Proceedings of the American Gas Association Conference* 403–405.

Gatley, D. P. 1988. Simplified life cycle costing of chilled water plants. *Heating, Piping & Air Conditioning* 60 (9): 55–68.

Geitner, F., and Galster, D. 2000. Using life-cycle costing tools. *Chemical Engineering* 107 (2): 80–86.

Gertz, M. 1997. Life cycle costing of gearboxes. *South African Mechanical Engineer* 47 (7): 21–22.

Gibbs, D. J. L., and King, R. J. 1989. Life cycle costing in the design of naval equipment. *Proceedings of the Conference on Reliability* 2A/5/1–2A/5/8.

Gibson, W. L., and Hartig, J. H. 1997. From life-cycle assessment to full-cost accounting: An evolving common language for cross-functional teams. *SAE Special Publications on Design for Environmentally Safe Automotive Products and Processes* 1263:69–73.

Girsch, G., Heyder, R., Kumpfmuller, N., and Belz, R. 2005. Comparing the life-cycle costs of standard and head-hardened rail. *Railway Gazette International* 161 (9): 549–551.

Gluch, P., and Baumann, H. 2004. The life cycle costing (LCC) approach: A conceptual discussion of its usefulness for environmental decision-making. *Building and Environment* 39 (5): 571–580.

Goble, W. M., and Paul, B. O. 1995. Life cycle cost estimating. *Chemical Processing* 58 (6): 5–6.

Goedecke, M., Therdthianwong, S., and Gheewala, S. H. 2007. Life cycle cost analysis of alternative vehicles and fuels in Thailand. *Energy Policy* 35 (6): 3236–3246.

Goel, P. S., and Singh, N. 1998. A framework for integrating quality, reliability, and durability in product design with life-cycle cost considerations. *Quality Engineering* 10 (2): 267–281.

Govil, K. K. 1992. A simple model for life cycle cost vs. maintainability function. *Microelectronics and Reliability* 32 (1–2): 269–270.

Graham, M. 2005. Life cycle management—Lowest cost per tonne: The heart of Joy's China strategy. *Coal International* 253 (5): 192–195.

Graham, S. 2007. Low life-cycle cost centrifugal pumps for utility applications. *World Pumps* 484:30–33.

Gransberg, D. D., and Diekmann, J. 2004. Quantifying pavement life cycle cost inflation uncertainty. *Proceedings of the AACE International Annual Meeting* RISK.08.1–RISK.08.11.

Gransberg, D. D., and Molenaar, K. R. 2004. Life-cycle cost award algorithms for design/build highway pavement projects. *Journal of Infrastructure Systems* 10 (4): 167–175.

Gratsos, G. A., and Zachariadis, P. 2005. Life cycle cost of maintaining the effectiveness of a ship's structure and environmental impact of ship design parameters. *Proceedings of the Royal Institution of Naval Architects International Conference* 95–122.

Gray, C. G., and Aase, B. K. 1994. Using simulation to assess manning, skills and logistics requirements for high productivity and low life cycle cost. *Proceedings of the European Production Operation Conference* 189–196.

Grayson, P. E., and Law, W. 1999. New instrumentation technology offers reduced life-cycle cost for maintaining geotechnical structures and other infrastructure. *Geotechnical News* 17 (2): 33–36.

Green, A. 1999. Life cycle costing for batteries in telecom applications. *Proceedings of the Twentieth International Telecommunications Energy Conference* 1–7.

Green, M. A. 1999. The future of minimal manning and its effects on the acquisition and life-cycle costs of major Coast Guard cutters. *Marine Technology* 36 (1): 55–59.

Greene, L. E. 1991. Life cycle cost (LCC) milestones. *Proceedings of the IEEE National Aerospace and Electronics Conference* 1197–1200.

Greene, L. E., and Shaw, B. L. 1990. The steps for successful life cycle cost analysis. *Proceedings of the IEEE National Aerospace and Electronics Conference* 1209–1216.

Gregorski, T. 2004. Interested in saving money? Control your life cycle costs. *Water and Wastes Digest* 44 (2): 10, 21.

Gregory, P. C., Donovan, K. S., and Spooner, O. R. 1993. Radioactive materials transportation life-cycle cost. *Transactions of the American Nuclear Society* 68 (3): 61–62.

Greyvenstein, G. P., and Van Niekerk, W. M. K. 1999. Life-cycle cost comparison between heat pumps and solar water heaters for the heating of domestic water in South Africa. *Journal of Energy in Southern Africa* 10 (3): 86–91.

Gurung, N., and Mahendran, M. 2002. Comparative life cycle costs for new steel portal frame building systems. *Building Research and Information* 30 (1): 35–46.

Gustafsson, S., and Karlsson, B. G. 1988. Why is life-cycle costing important when retrofitting buildings? *International Journal of Energy Research* 12 (2): 233–242.

———. 1989. Life cycle cost minimization considering retrofits in multi-family residences. *Energy and Buildings* 14 (1): 9–17.

Gustafsson, S. et al. 1991. Window retrofits: Interaction and life-cycle costing. *Applied Energy* 39 (1): 21–29.

Gustavsson, J. 2002. Software program that calculates the life cycle cost of air filters. *Filtration and Separation* 39 (9): 22–26.

Haas, R. C. G., Tighe, S. L., and Falls, L. C. 2005. Life-cycle cost analysis protocol for infrastructure assets. *Proceedings of the Annual Canadian Society for Civil Engineering Conference* FR-128-1–FR-128-12.

Hackney, J., and de Neufville, R. 2001. Life cycle model of alternative fuel vehicles: Emissions, energy, and cost trade-offs. *Transportation Research Part A: Policy and Practice* 35 (3): 243–266.

Hagen, C. J., and Brouwers, G. 1994. Reducing software life-cycle costs by developing configurable software. *Proceedings of the IEEE National Aerospace and Electronics Conference* 1182–1187.

Hall, M. J. 1994. Life cycle cost implementation in electronics. *Proceedings of the 2nd International Conference on Concurrent Engineering and Electronic Design Automation* 189–194.

Hall, S. C. 2004. Nuclear plant life cycle cost analysis considerations. *Proceedings of the 2004 International Congress on Advances in Nuclear Power Plants* 790–799.

Hamel, R. C. 1991. Managing life cycle costs. *Proceedings of the Test Engineering Conference* 177–180.

Hamer, P. S. et al. 1996. Energy-efficient induction motors performance characteristics and life-cycle cost comparison for centrifugal loads. *Proceedings of the Annual Petroleum and Chemical Industry Conference* 209–217.

———. 1997. Energy-efficient induction motors performance characteristics and life-cycle cost comparisons for centrifugal loads. *IEEE Transactions on Industry Applications* 33 (5): 1312–1320.

Harding, T. B. 1996. Life cycle value/cost decision making. *Proceedings of the SPE International Petroleum Conference & Exhibition* 143–152.

Hartmann, A. et al. 2000. Life cycle cost modeling of continuous fiber reinforced thermoplastics. *Proceedings of the 45th International SAMPE Symposium and Exhibition* 1081–1091.

Hasan, A. 1999. Optimizing insulation thickness for buildings using life cycle cost. *Applied Energy* 63 (2): 115–124.

Hawk, H. 2003. *Bridge life cycle cost analysis.* Washington, D.C.: Transportation Research Board.

Hayek, M. E., van Voorthuysen, E., and Kelly, D. W. 2005. Optimizing life cycle cost of complex machinery with rotable modules using simulation. *Journal of Quality in Maintenance Engineering* 11 (4): 333–347.

Hegazy, T. et al. 2004. Bridge deck management system with integrated life-cycle cost optimization. *Transportation Research Record* 1866:44–50.

Hegde, G. G. 1994. Life cycle cost: A model and applications. *IIE Transactions* 26 (6): 56–62.

Hellgren, J. 2007. Life cycle cost analysis of a car, a city bus and an intercity bus power train for year 2005 and 2020. *Energy Policy* 35 (1): 39–49.

Hellmann, D. 1998. Reduction of life-cycle-costs by early diagnosis of failures. *Proceedings of the International Conference on Pumps and Fans* 94–102.

Hennecke, F. W. 1999. Life cycle costs of pumps in chemical industry. *Chemical Engineering and Processing* 38 (4–6): 511–516.

———. 2006. A comparative study of pump life cycle costs. *Paper Technology* 47 (7): 20–27.

Henninger, A. 1993. Reducing weapons systems' life cycle costs with simulation modeling. *Computers and Industrial Engineering* 25:183–185.

Hodowanec, M. M. 1998. Evaluation of anti-friction bearing lubrication methods on motor life cycle cost. *Proceedings of the IEEE Annual Pulp and Paper Industry Technical Conference* 196–201.

———. 1999. Evaluation of antifriction bearing lubrication methods on motor life-cycle cost. *IEEE Transactions on Industry Applications* 35 (6): 1247–1251.

Hoff, J. L. 2001. Roofing and life cycle cost. *Buildings: Cedar Rapids* 95 (5): 74–75.

Holt, W. L. 2001. Utility tracks transformer inventory and life-cycle costs. *Transmission and Distribution World* 53 (2): 70–74.

Hombach, W. G. 1995. Evaluation of environmental management cost estimating capabilities for the subject area "life-cycle economics for radioactive waste management and environmental remediation." *Proceedings of the International Conference on Radioactive Waste Management and Environmental Remediation* 181–185.

Hong, T., Han, S., and Lee, S. 2007. Simulation-based determination of optimal life-cycle cost for FRP bridge deck panels. *Automation in Construction* 16 (2): 140–152.

Hossain, A. L. F. M., Bradley, P. J., Walker, J., and Wingerter, R. G. 1992. Life cycle cost management in a multiple supplier environment—an implementation case study. *Proceedings of the IEEEE Conference on Discovering a New World of Communications* 1772–1778.

Houshyar, A. 2005. Reliability and maintainability of machinery and equipment, part 2: Benchmarking, life-cycle cost, and predictive maintenance. *International Journal of Modeling and Simulation* 25 (1): 1–11.

Howard, R. J. 1991. Road life cycle costing. *Proceedings of the Institution of Engineers Australia National Conference on Effective Management of Assets and Environment* 55–59.

Howarth, J. 2004. Life preserver (pump–life cycle costing). *Engineer* 293 (7663): 51.

Hu, K. X., Knecht, T., Yeh, C. P., Mui, G., and Wyatt, K. W. 1997. A total product life cycle profile approach to reliability analysis for low cost crystal oscillators. *Proceedings of the Pacific Rim/ASME International Intersociety Electronic and Photonic Packaging Conference* 555–560.

Hu, Z. Y. et al. 2004. Net energy, CO2 emission, and life-cycle cost assessment of cassava-based ethanol as an alternative automotive fuel in China. *Applied Energy* 78 (3): 247–256.

Hunkeler, D., Lichtenvort, K., and Rebitzer, G., eds. 2008. *Environmental life cycle costing*. Boca Raton, FL: CRC Press.

Hutton, R. 1994. Condition monitoring and its contribution to life cycle costs. *IEEE Colloquium on Life Cycle Costing and Business Plan Digest* 13:6/1–6/4.

Hwang, H. 1999. A FMS performance analysis model based on system availability and life cycle cost. *Journal of Engineering Valuation and Cost Analysis* 2 (2): 143–149.

Hyong-Bok, K., and Kyong-Min, K. 2000. Generating water-distribution and sewer network alternatives using models, value engineering, life cycle costing and GIS. *Proceedings of the Annual Conference of the Urban and Regional Information Systems Association* 668–680.

Ibrahim, M. Y., and Brack, C. 2004. New concept and implementation of intercontinental flexible training of terotechnology and life cycle costs. *Proceedings of the IEEE International Conference on Industrial Technology* 224–229.

In, H. et al. 2006. A quality-based cost estimation model for the product line life cycle. *Communications of the ACM* 49 (12): 85–88.

Jackson, A. M. 1994. Emerging standards reduce product life-cycle costs. *Proceedings of the IEEE Systems Readiness Technology Conference* 131–138.

Janz, D., Sihn, W., and Warnecke, H. J. 2005. Product redesign using value-oriented life cycle costing. *CIRP Annals—Manufacturing Technology* 54 (1): 9–12.

Jeong, K. S., and Oh, B. S. 2002. Fuel economy and life-cycle cost analysis of a fuel cell hybrid vehicle. *Journal of Power Sources* 105 (1): 58–65.

Jha, N. K., and Litkouhi, B. 2001. Optimal life cycle cost analysis and design of thermal systems. *Proceedings of the ASEE Annual Conference* 7671–7683.

Jiang, M. et al. 2000. Optimal life-cycle costing with partial observability. *Journal of Infrastructure Systems* 6 (2): 56–66.

Jiang, R., Zhang, W. J., and Ji, P. 2003. Required characteristics of statistical distribution models for life cycle cost estimation. *International Journal of Production Economics* 83 (2): 185–194.

———. 2004. Selecting the best alternative based on life-cycle cost distributions of alternatives. *International Journal of Production Economics* 89 (1): 69–75.

Jin, N. H., Chryssanthopoulos, M. K., and Parke, G. A. R. 2005. Bridge management using principles of whole life costing and life cycle assessment subject to uncertainty. *Proceedings of the 5th International Conference on Bridge Management* 426–432.

Johnson, B., Powell, T., and Queiroz, C. 1998. Economic analysis of bridge rehabilitation options considering life cycle costs. *Transportation Research Record* 1624:8–15.

Johnson, K. 1999. Have we forgotten about true life cycle cost in electronics assembly? (Or how to be a hero in your boss's eyes). *Proceedings of the National Electronic Packaging and Production Conference* 659–672.

Johnson, V. S. 1990. Minimizing life cycle cost for subsonic commercial aircraft. *Journal of Aircraft* 27 (2): 139–145.

Jones, C. 1994. Life cycle cost models. *IEE Colloquium Digest* 013:2/1–2/6.

Jones, P. 1989. Naval life cycle costing—still a black art to industry? *Proceedings of the Conference on Military Computers Systems and Software* 255–260.

Jurges, G. F. 1999. Performance based simulation modeling quantifies aircraft carrier life cycle cost and readiness. *Naval Engineers Journal* 111 (1): 27–38.

Jyrkama, M. I., and Pandey, M. D. 2006. The impact of aging on life cycle cost: Techniques for analysis and optimization. *Proceedings of the Annual Conference of the Canadian Nuclear Society* 16–17.

Kage, I. et al. 2005. Minimum maintenance steel plates and their application technologies for bridge—Life cycle cost reduction technologies with environmental safeguards for preserving social infrastructure assets. *JFE Technical Report: Steel Plates* 5:37–44.

Kaito, K., Abe, M., Koide, Y., and Fujino, Y. 2001. Bridge management strategy for a steel plate girder bridge based on minimum total life cycle cost. *Proceedings of SPIE Conference* 194–202.

Kaminski, M. L. 1993. Cost-effective life-cycle design of fatigue costive structural components. *Proceedings of the 1st Joint Conference on Marine Safety and Environment Ship Production* 645–657.

Kaminski, M. L., Boonstra, H., Salza, P., and Wittenberg, L. 1993. Cost effective life-cycle design of semisubmersibles based on probabilistic fatigue calculations. *Proceedings of the International Conference on Offshore Mechanics and Arctic Engineering* 369–378.

Kannan, R., Leong, K. C., Osman, R., and Tso, C. P. 2005. Gas-fired combined cycle plant in Singapore: Energy use, GWP, and cost-a life cycle approach. *Energy Conversion and Management* 46 (13–14): 2145–2157.

Kapoor, L. M. 1990. Determining life cycle costs of a work measurement system (WMS). *Proceedings of the IEEE National Aerospace and Electronics Conference* 995–1000.

Karlsson, D. et al. 1997. Reliability and life cycle cost estimates of 400 kV substation layouts. *IEEE Transactions on Power Delivery* 12 (4): 1486–1492.

Karyagina, M., Wong, W., and Vlacic, L. 1998. Life cycle cost modeling using marked point processes. *Reliability Engineering and System Safety* 59 (3): 291–298.

Keene, S. J., and Keene, K. C. 1993. Reducing the life cycle cost of software through concurrent engineering. *Proceedings of the Annual Reliability and Maintainability Symposium* 305–310.

Khanduri, A. C., Bedard, C., and Alkass, S. 1993. Life cycle costing of office buildings at the preliminary design stage. *Proceedings of the 5th International Conference on Civil and Structural Engineering Computing* 1–8.

———. 1996. Assessing office building life cycle costs at preliminary design stage. *Structural Engineering Review* 8 (2–3): 105–114.

Kiessling, R. 1990. Reduced life cycle costs, neglected arguments for stainless steel. *Steel Times* 218 (1): 29–30, 32.

King, S. A., Jain, A., and Hart, G. C. 2001. Life-cycle cost analysis of supplemental damping. *Structural Design of Tall Buildings* 10 (5): 351–360.

Kirk, S. J. 1996. Life cycle costing reveals masonry's long-term value. *Masonry Construction* December: 555–557.

Kirk, S. J., and Dell'Isola, A. J. 1995. *Life cycle costing for design professionals.* New York: McGraw–Hill.

Kirkham, R. J. 2005. Re-engineering the whole life cycle costing process. *Construction Management and Economics* 23 (1): 9–14.

Kirkham, R. J. et al. 2002. Probability distributions of facilities management costs for whole life cycle costing in acute care NHS hospital buildings. *Construction Management and Economics* 20 (3): 251–261.

Kleyner, A., Sandborn, P., and Boyle, J. 2004. Minimization of life cycle costs through optimization of the validation program: A test sample size and warranty cost approach. *Proceedings of the Annual Reliability and Maintainability Symposium* 553–558.

Knight, R. S. 1989. Life cycle costing: An industry view on the way ahead. *Proceedings of the Conference on Military Computers Systems and Software* 261–266.

Kohlhase, N. 2001. Cutting life cycle costs. *Hydrocarbon Engineering* 6 (2): 83–85.

Koner, P. K., Dutta, V., and Chopra, K. L. 2000. A comparative life cycle energy cost analysis of photovoltaic and fuel generator for load shedding application. *Solar Energy Materials and Solar Cells* 60 (4): 309–322.

Kong, J. S., and Frangopol, D. M. 2003. Life-cycle reliability-based maintenance cost optimization of deteriorating structures with emphasis on bridges. *Journal of Structural Engineering* 129 (6): 818–828.

———. 2004. Cost-reliability interaction in life-cycle cost optimization of deteriorating structures. *Journal of Structural Engineering* 130 (11): 1704–1712.

Konig, N., and Bayley, C. 2001. Euro-interlocking promises lower life-cycle costs. *Railway Gazette International* 157 (10): 683–686.

Kopscick, G. 2002. Life cycle costs. *Hydrocarbon Engineering* 7 (2): 59–64.

Kostic, S., and Pendic, Z. 1988. System design and development-a life-cycle cost approach. *Proceedings of the 7th Symposium on Reliability in Electronics* 246–253.

Kroon, D. H. 2006. Life-cycle cost comparisons of corrosion protection methods for ductile iron pipe. *Materials Performance* 45 (5): 44–48.

Kumar, G. H., and Govindaraj, S. R. 2001. Energy efficient motors and life cycle costing analysis. *IEEMA Journal* 21 (8): 26–34.

Kumaran, K. D., Ong, S. K., Tan, R. B. H., and Nee, A. Y. C. 2001. Tool to incorporate environmental costs into life cycle assessment. *Proceedings of the SPIE Conference* 124–134

Laitinen, M., Heikkinen, U., and Launonen, U. 2007. Minimizing life cycle costs with modern consistency transmitters. *Appita Journal* 60 (3): 191–195.

Lam, J. C. 1993. Energy-efficient measures and life cycle costing of a residential building in Hong Kong. *Architectural Science Review* 36 (4): 157–162.

Lam, J. C., and Chan, A. L. S. 1995. Life-cycle costing of energy-efficient measures for commercial buildings. *Architectural Science Review* 38 (3): 125–131.

Lam, J. C., and Chan, W. W. 2001. Life cycle and green cost analysis of energy-efficient lighting for hotels. *Architectural Science Review* 44 (2): 135–138.

———. 2001. Life cycle energy cost analysis of heat pump application for hotel swimming pools. *Energy Conversion and Management* 42 (11): 1299–1306.

Landamore, M., Birmingham, R., and Downie, M. 2007. Establishing the economic and environmental life-cycle costs of marine systems: A case study from the recreational craft sector. *Marine Technology* 44 (2): 106–117.

Lansdowne, Z. F. 1994. Built-in test factors in a life cycle cost model. *Reliability Engineering and System Safety* 43 (3): 325–330.

Larsen, C., Szaro, J., and Wilson, W. 2004. An alternative approach to PV system life cycle cost analysis. *Proceedings of the International Solar Energy Conference* 415–420.

Larsen, C., Szaro, J., Wilson, W., and Lynn, K. 2005. An alternative approach to PV system life cycle cost analysis (PV LCC): Phase II. *Proceedings of the International Solar Energy Conference* 447–452.

Lassen, T., and Syvertsen, K. 1996. Fatigue reliability and life cycle cost analysis of mooring chains. *Proceedings of the International Offshore and Polar Engineering Conference* 418–422.

———. 1997. Fatigue reliability and life-cycle cost analysis for mooring chains. *International Journal of Offshore and Polar Engineering* 7 (2): 135–140.

Lee, D. B. 2002. Fundamentals of life-cycle cost analysis. *Transportation Research Record* 1812:203–210.

Lee, K., Cho, H., and Cha, C. S. 2006. Life-cycle cost-effective optimum design of steel bridges considering environmental stressors. *Engineering Structures* 28 (9): 1252–1265.

Lee, K., Cho, H., and Choi, Y. 2004. Life-cycle cost-effective optimum design of steel bridges. *Journal of Constructional Steel Research* 60 (11): 1585–1613.

Lee, K., Cho, H., Lim, J., and Park, K. 2003. Life-cycle cost effective optimal seismic design for continuous PSC bridges. *Proceedings of the Conference on the Life-Cycle Performance of Deteriorating Structures, Assessment, Design and Management* 247–262.

Lee, Y., and Chang, L. 2003. Rehabilitation decision analysis and life-cycle costing of the infrastructure system. *Proceedings of the Construction Research Congress* 691–701.

Leech, D. J., and Etemad, F. 1989. Life cycle cost prediction. *Proceedings of the Fifth National Conference on Production Research* 128–133.

Leeming, M. B. 1993. Application of life cycle costing to bridges. In *Bridge management*, 574–583. London: Thomas Telford Services Ltd.

Lefton, S. A., Besuner, P. M., and Grimsrud, G. P. 1995. Managing utility power plant assets to economically optimize power plant cycling costs, life, and reliability. *Proceedings of the 4th IEEE Conference on Control Applications* 195–208.

Leiter, A. J., and Wowak, W. E. 1989. Models developed for the total system life-cycle cost analysis, *Transactions of the American Nuclear Society* 60:158–159.

Leitner, G. F., and Leitner, W. 1994. Life cycle and present worth cost concepts, applicable for large desalting plants? *Desalination* 97 (1–3): 291–300.

Leppert, S. M., and Allen, A. D. 1995. Conductor life cycle cost analysis. *Proceedings of the IEEE Rural Electric Power Conference* C2.1–C2.8.

Lesnoski, T. M., Life cycle cost (LCC) estimating for large management information system (MIS) software development projects. *Proceedings of the IEEE National Aerospace and Electronics Conference* 937–943.

Liang, Q. W., and Song, B. W. 2005. Combination of latent root regression and fuzzy-gray theory which was used for the life cycle cost modeling of weapon system. *Journal of Information and Computation Science* 2 (2): 273–282.

———. 2005. Fuzzy regression model of torpedo life cycle cost based on the fuzzy output. *Journal of Information and Computation Science* 2 (3): 517–522.

Lien, Y. C., and Narula, R. G. 1989. Fossil life cycle management: A key to cost competitiveness. *Proceedings of the International Conference for the Power Generation Industries* 735–747.

Liosis, A. C. 2001. A prime contractor's perspective on total ATS development and life cycle cost (LCC) support responsibility. *Proceedings of the IEEE Systems Readiness Technology Conference* 313–318.

Liu, M., Burns, S. A., and Wen, Y. K. 2003. Optimal seismic design of steel frame buildings based on life cycle cost considerations. *Earthquake Engineering and Structural Dynamics* 32 (9): 1313–1332.

———. 2004. Multiobjective optimization for life cycle cost oriented seismic design of steel moment frame structures. *Proceedings of the 2004 Structures Congress* 1391–1394.

Liu, M., and Frangopol, D. 2005. Multiobjective maintenance planning optimization for deteriorating bridges considering condition, safety, and life-cycle cost. *Journal of Structural Engineering* 131 (5): 833–842.

Liu, M., Wen, Y. K., and Burns, S. A. 2004. Life cycle cost oriented seismic design optimization of steel moment frame structures with risk-taking preference. *Engineering Structures* 26 (10): 1407–1421.

Livingston, R. A. 1990. Service life prediction and life-cycle costing for materials damage as a result of acid deposition. *ASTM Special Technical Publication* 1098: 40–56.

Lutz, J. et al. 2006. Life-cycle cost analysis of energy efficiency design options for residential furnaces and boilers. *Energy* 31 (2–3): 311–329.

Lynn, K. et al. 2006. A review of PV system performance and life-cycle costs for the Sunsmart schools program. *Proceedings of the International Solar Energy Conference* 153–156.

Macfarlane, M. S., and Mackey, P. A. 1998. Monobores—Making a difference to the life cycle cost of a development. *Proceedings of the Asia Pacific Oil & Gas Conference* 137–143.

———. 1999. Monobores improve life-cycle cost. *Journal of Petroleum Technology* 51 (2): 69–70.

Mack, J. 1999. State DOTs update life cycle cost analysis. *Better Roads* October: 25–28.

Maharsia, R. R., and Jerro, H. D. 2002. Investigation of the manufacturability of smart composite piping structures using life cycle cost modeling and uncertainty analysis. *Proceedings of the ASME Engineering Technology Conference on Energy* 153–160.

Malhotra, V., Khan, J. R., Lear, W. E., and Sherif, S. A. 2005. Life cycle cost analysis of a novel cooling and power gas turbine engine. *Proceedings of the ASME International Mechanical Engineering Congress* 79–92.

Malik, M. A. K., and Kolodchak, P. 1990. Cost-reliability relationship in life cycle. *Proceedings of the International Industrial Engineering Conference* 581–586.

Malkki, H., Enwald, P., and Toivonen, J. 1991. Experience of transferring life cycle costing to manufacturing industry. *Proceedings of the ASME International Power Generation Conference* 1–8.

———. 1991. Life cycle costing and condition monitoring applied in a pump system. *Proceedings of the 12th Annual Symposium of the Society of Reliability Engineers* 300–310.

Markeset, T., and Kumar, U. 2001. R & M and risk-analysis tools in product design, to reduce life-cycle cost and improve attractiveness. *Proceedings of the Annual Reliability and Maintainability Symposium* 116–122.

Markow, M. J. 1989. Life-cycle cost evaluations of payment construction, rehabilitation, and maintenance. *Proceedings of the Annual Conference of the Urban and Regional Information Systems Association* 13–17.

Marr, W. W., and Walsh, W. J. 1992. Life-cycle cost evaluations of electric/hybrid vehicles. *Energy Conversion and Management* 33 (9): 849–853.

Martin, L. 1996. Radiation tolerant computer design to meet customer interface requirements for miniature inertial measurement unit (MIMU) space applications emphasizing low life cycle costs. *Proceedings of the Annual Rocky Mountain Guidance and Control Conference* 359–371.

Martin, T., Michel, N., and Potter, N. 2000. Road database needs for a network life-cycle costing analysis. *Proceedings of the Conference of the Australian Road Research Board* 57–72.

Martin, T., Potter, N., and Michel, N. 2001. Road database needs for a network life-cycle costing analysis. *Road and Transport Research* 10 (4): 42–53.

Martin, T. C. 1998. Road network asset management using pavement life-cycle costing modeling. *Proceedings of the Conference of the Australian Road Research Board* 187–203.

Martin, T. C., and Taylor, S. Y. 1994. Life-cycle costing: Prediction of pavement behavior. *Proceedings of the Conference of the Australian Road Research Board* 187–206.

Marx, W. J., Mavris, D. N., and Schrage, D. P. 1996. Effects of alternative wing structural concepts on high-speed civil transport life cycle costs. *Proceedings of the AIAA/ ASME/ASCE/AHS/ASC Structures, Structural Dynamics & Materials Conference* 562–582.

Maxwell, D. 1993. Improving life cycle costs for industrial plants. *World Cement* 24 (6): 33–34.

McArthur, C. J., and Snyder, H. M. 1989. Life cycle cost-the logistics support analysis connection. *Proceedings of the IEEE National Aerospace and Electronics Conference* 1206–1209.

McCormac, D. E. et al. 1990. Economic implications of space-reliability EEE parts and program life cycle costs. *Proceedings of the International Symposium on Reliability and Maintainability* 398–403.

McDonagh, J. F., and Hopman, J. H. 1991. Design criteria for minimum life cycle costs of university buildings. *Proceedings of the National Conference of the Institution of Engineers of Australia* 21–26.

McDowall, J. 2001. Battery life considerations in energy storage applications and their effect on life cycle costing. *Proceedings of the Power Engineering Society Summer Meeting* 452–455.

McKeever, B. et al. 1998. Life cycle cost-benefit model for road weather information systems. *Transportation Research Record* 1627:41–48.

McManus, K. J. et al. 1998. Deterioration models and life cycle costing, for local street concrete pavements, within the city of Stonnington. *Proceedings of the Conference of the Australian Road Research Board* 34–48.

McNichols, G. R. 1988. Life cycle cost—art or science? *Proceedings of the IEEE National Aerospace and Electronics Conference* 1428–1433.

Meiarashi, S., Nishizaki, I., and Kishima, T. 2002. Life-cycle cost of all-composite suspension bridge. *Journal of Composites for Construction* 6 (4): 206–214.

Meisl, C. J. 1988. Life-cycle cost considerations for launch vehicle liquid propellant rocket engine. *Journal of Propulsion and Power* 4 (2): 118–126.

———. 1988. Life-cycle methodology for space station propulsion system. *Journal of Propulsion and Power* 4 (2): 111–117.

Merkel, T., and Tione, R. 2005. Fleet management—Life cycle cost-based maintenance supported by advanced brake systems. *ZEV Rail Glasers Annalen* 129:84–95.

Mevellec, P., and Perry, N. 2006. Whole life-cycle costs—A new approach. *International Journal of Product Lifecycle Management* 1 (4): 400–414.

Meyer, J. J. 1990. Look back in time to verify life cycle cost analyses. *Proceedings of the International Conference on Pipeline Design and Installation* 630–638.

Migliaccio, G. C. et al. 2006. Life-cycle cost analysis for selection of energy-efficient building components in lodging facilities. *Proceedings of the Architectural Engineering National Conference* 54–64.

Miller, J., and Miller, B. 2004. Life cycle cost reduction for reciprocating slurry pump stations. *Proceedings of the Hydrotransport 16th International Conference* 163–175.

Mills, D. J. 1994. Elements of total life cycle costing. *Proceedings of the Australasian Institute of Mining & Metallurgy, Conference and Workshop* 95–99.

Millward, D. G. 1996. Life-cycle cost trade studies for hardness assurance. *IEEE Transactions on Nuclear Science* 43 (6): 3133–3138.

Mohammadi, J., Guralnick, S. A., and Yan, L. 1995. Incorporating life-cycle costs in highway-bridge planning and design. *Journal of Transportation Engineering* 121 (5): 417–424.

Monga, A., Zuo, M. J., and Toogood, R. 1995. System design for minimal life cycle cost. *Proceedings of the 4th Industrial Engineering Research Conference* 335–341.

———. 1995. System design with deteriorative components for minimal life cycle costs. *Proceedings of the IEEE International Conference on Systems, Man and Cybernetics. Intelligent Systems for the 21st Century* 1843–1848.

Morgan, S. M. et al. 2001. Study of noise barrier life-cycle costing. *Journal of Transportation Engineering* 127 (3): 230–236.

Morris, J. 1998. A methodology for evaluating pintle system life cycle costs. *Proceedings of the 36th Annual SAFE Symposium* 516–523.

Morton, B. S., Visser, A. T., and Horak, E. A. 2006. Life cycle cost analysis of the Gauteng to Durban freight corridor: Introduction to study. *Proceedings of the 25th Annual Southern African Transport Conference* 2006:475–484.

Munteanu, R. 1994. Silicone rubber insulators reduce life cycle costs. *Transmission and Distribution International* 5 (1): 18, 21–22, 25.

Najafi, M., and Kim, K. 2004. Life-cycle-cost comparison of trenchless and conventional open-cut pipeline construction projects. *Proceedings of the ASCE Pipeline Division Specialty Congress on Pipeline Engineering and Construction* 635–640.

Nakabayashi, M., Takano, T., Temma, K., and Iyoda, I. 2000. Optimal design method for BTB based on reliability and life cycle cost evaluation. *Proceedings of the International Conference on Power System Technology* 739–744.

Nam, S. H. 2003. The optimal policy of quality improvement under expected warranty costs and product life cycle. *Proceedings of the Annual Meeting of the Decision Sciences Institute* 1735–1740.

Nassar, K., Beliveau, Y., and Ellis, M. 1997. Financing and life cycle cost issues versus quality in residential construction. *Proceedings of the ASCE Construction Congress* 832–840.

Neely, E. S., and Neathammer, R. D. 1989. Building life cycle costs in the United States Army. *Proceedings of the Third Conference on Human–Computer Interaction* 147–154.

———. 1989. Computerized life-cycle cost systems in the Army. *Journal of Computing in Civil Engineering* 3 (1): 93–104.

———. 1989. Life cycle costs in the army. *Proceedings of the 6th Conference on Computing in Civil Engineering* 653–660.

———. 1989. Teaching undergraduates and graduates life cycle cost procedures at Penn State University through research. *Proceedings of the SCS Western Multiconference* 106–109.

———. 1991. Life-cycle maintenance costs by facility use. *Journal of Construction Engineering and Management* 117 (2): 310–320.

Nel, J. J., Swart, P. H., and Case, M. J. 1996. Life cycle cost comparison of laser modulator topologies. *Proceedings of the IEEE Power Modulator Symposium* 85–88.

Nesbitt, B. 2001. Intelligent pump units and life cycle costs. *World Pumps* 416:32–36.

Neumann, S. B., and Fenton, D. L. 1992. Life-cycle cost analysis applied to selection of compression equipment for industrial refrigeration systems. *ASHRAE Transactions* 98 (2): 148–155.

Newton, L., and Christian, J. 2000. Design of a data management system for life cycle cost analysis. *Proceedings of the CSCE Annual Conference* 342–348.

Nickerson, R. L. 1995. Life-cycle cost analysis for highway bridges. *Proceedings of the CSCE Structures Congress* 676–677.

Nilsson, J., and Bertling, L. 2007. Maintenance management of wind power systems using condition monitoring systems—Life cycle cost analysis for two case studies. *IEEE Transactions on Energy Conversion* 22 (1): 223–229.

Niwa, M., Kato, T., and Suzuoki, Y. 2005. Life-cycle-cost evaluation of degradation diagnosis for cables. *Proceedings of the International Symposium on Electrical Insulating Materials* 737–740.

Nowak, E. 2003. Product life cycle cost management. *Management* 7 (1): 157–162.

O'Malley, C. M. 1992. Life cycle costing for reality in project design evaluation. *Power Technology International* 27–28, 30–31.

Oman, H. 2002. Performance, life-cycle cost, and emissions of fuel cells. *IEEE Aerospace and Electronic Systems Magazine* 17 (9): 33–37.

O'Neil, G. et al. 1995. A proposed framework for the application of probabilistic geotechnics in the optimization of pipeline life cycle cost. *Proceedings of the International Conference on Offshore Mechanics and Arctic Engineering* 97–106.

Osman, H. 2005. Risk-based life-cycle cost analysis of privatized infrastructure. *Transportation Research Record* 1924:192–196.

Ozbay, K. et al. 2004. Life-cycle cost analysis: State of the practice versus state of the art. *Transportation Research Record* 1864:62–70.

Papagiannakis, T., and Delwar, M. 2001. Computer model for life-cycle cost analysis of roadway pavements. *Journal of Computing in Civil Engineering* 15 (2): 152–156.

Pappas, C. P. 1992. Fiber in the loop (FITL) life cycle cost analysis. *Proceedings of the SPIE Conference* 1578:110–114.

Park, J. H., and Seo, K. K. 2004. Incorporating life-cycle cost into early product development. *Journal of Engineering Manufacture* 218 (9): 1059–1066.

Park, J. H., Seo, K. K., Wallace, D., and Lee, K. I. 2002. Approximate product life cycle costing method for the conceptual product design. *CIRP Annals—Manufacturing Technology* 51 (1): 421–424.

Paterson, R. 1995. Minimizing the life cycle costs attributed to boiler tubing in fossil-fueled plants. *Proceedings of the American Power Conference* 1787–1797.

Paul, B. O. 1994. Life cycle costing. *Chemical Processing* 57 (12): 79–83.

Perera, H. S. C., Nagarur, N., and Tabucanon, M. T. 1999. Component part standardization: A way to reduce the life-cycle costs of products. *International Journal of Production Economics* 60-61:109–116.

Petersen, K. E., Rasmussen, B., and Jensen, P. H. 1989. Reliability analysis in life cycle cost estimation for small wind turbines. *Proceedings of the 10th Annual Symposium of the Society of Reliability Engineers* 90–98.

Pfohl, M. C. 1999. Prototype-based life cycle costing in the R&D. *Proceedings of the Portland International Conference on Management of Engineering and Technology* 445.

Pham, H. 1996. Software cost model with imperfect debugging, random life cycle and penalty cost. *International Journal of Systems Science* 27 (5): 455–463.

Phillips, R., and Brown, B. 1999. Life cycle cost of military displays. *Proceedings of the International Society for Optical Engineering Conference on Cockpit Displays* 138–147.

Pierce, P. 1997. Covered bridges—Life-cycle cost advantages. *Proceedings of the Structures Congress* 238–242.

Pigoski, T. M. 1994. Reducing life cycle costs. *Managing System Development* 14 (12): 5–7.

Plebani, S., Rosi, R., and Zanetta, L. 1996. Life cycle costs comparison and sensitivity analysis for multimedia networks. *IEE Colloquium on Optical and Hybrid Access Networks*, Digest no.1996/052 3/1–3/5.

Politano, D., and Frohlich, K. 2006. Calculation of stress-dependent life cycle costs of a substation subsystem—Demonstrated for controlled energization of unloaded power transformers. *IEEE Transactions on Power Delivery* 21 (4): 2032–2038.

Ponniah, J. E., and Kennepohl, G. J. 1996. Crack sealing in flexible pavements: A life-cycle cost analysis. *Transportation Research Record* 1529:86–94.

Pontarollo, J., Hooton, D., and Byer, P. 2000. Environmental life-cycle cost analysis of asphalt and concrete pavements. *Proceedings of the CSCE 6th Environmental Engineering Specialty Conference* 469–476.

Porter, J. 2000. The resurrection of life-cycle costing. *Chemical Processing* 63 (2): 74–79.

Prang, J. 1991. Controlling life-cycle costs through concurrent engineering. Work smarter not harder. *Proceedings of the ATE and Instrumentation Conference* 1–8.

Prasad, B. 1999. Model for optimizing performance based on reliability, life-cycle costs and other measurements. *Production Planning and Control* 10 (3): 286–300.

Proffitt, J. T. 1994. Life cycle costs of aircraft systems. *IEE Colloquium on Life Cycle Costing and the Business Plan Digest* 1994/103:1/1–1/3.

Quartier, F., and Wery, B. 1999. Can we compress the life cycle cost of complex real-time systems? *Proceedings of the Data Systems in Aerospace Conference* 349–352.

Rafiq, M. I., Chryssanthopoulos, M., and Onoufriou, T. 2005. Comparison of bridge management strategies using life-cycle cost analysis. *Proceedings of the 5th International Conference on Bridge Management, Inspection, Maintenance, Assessment, and Repair* 578–586.

Ramohalli, K., and Preiss, B. 1992. Quantitative figure-of-merit for space missions: Importance of the life-cycle costs. *Proceedings of the American Society of Mechanical Engineers Conference on Space Exploration Science and Technologies Research* 29–41.

Ravirala, V., and Grivas, D. 1995. State increment method of life-cycle cost analysis for highway management. *Journal of Infrastructure Systems* 1 (3): 151–159.

Ray, C. et al. 1999. Hazardous waste minimization through life cycle cost analysis at federal facilities. *Journal of the Air & Waste Management Association* 49 (1): 17–27.

Reich, M. C. 2005. Economic assessment of municipal waste management systems—Case studies using a combination of life cycle assessment (LCA) and life cycle costing (LCC). *Journal of Cleaner Production* 13 (3): 253–263.

Reigle, J. A., and Zaniewski, J. P. 2002. Risk-based life-cycle cost analysis for project-level pavement management. *Transportation Research Record* 1816:34–42.

Reinhard, G., and Hanna, J. 1993. Extension of TISSS test methodology from chip level to board level for improved transportability and decreased life-cycle costs. *Proceedings of the AUTOTESTCON '93 Conference* 173–179.

Renda-Tanali, I., and Hekimian, C. D. 2003. A simulation tool for life cycle costing of water supply infrastructure in seismically active zones. *International Journal of Emergency Management* 1 (4): 332–345.

Rensink, H. J. T., and Van Uden, M. E. J. 2004. Human factors engineering: An upfront engineering level of protection leading to improved human efficiency, better system performance and life cycle cost reductions: Part 1: The development of a human factors engineering strategy in petrochemical engineering and projects. *Proceedings of the SPE International Conference on Health, Safety and Environment in Oil and Gas Exploration and Production* 565–576.

Rey, F. J., Martin-Gil, J., Velasco, E., Perez, D., Varela, F., Palomar, J. M., and Dorado, M. P. 2004. Life cycle assessment and external environmental cost analysis of heat pumps. *Environmental Engineering Science* 21 (5): 591–605.

Ridilla, J. S., and Sathisan, S. K. 1998. A decision support tool to estimate life cycle costs of heavy-haul transport. *Proceedings of the 18th International Conference on High-Level Radioactive Waste Management* 818–820.

Riedel, T., Tiemann, N., Wahl, M. G., and Ambler, T. 1998. LCCA-life cycle cost analysis. *Proceedings of the IEEE Systems Readiness Technology Conference* 43–47.

Rigden, S. R., Burley, E., and Tajalli, S. M. A. 1995. Life cycle costing and the design of structures with particular reference to bridges. *Proceedings of the Institution of Civil Engineers, Municipal Engineer* 109 (4): 284–286.

Riggs, J. L., and Jones, D. 1990. Flowgraph representation of life cycle cost methodology—A new perspective for project managers. *IEEE Transactions on Engineering Management* 37 (2): 147–152.

Rinck, C. A. 1995. Medium voltage breaker rehabilitation: A life-cycle cost analysis of replacement, retrofit and interrupter technology options. *Proceedings of the International Conference on Hydropower–Waterpower* 1156–1161.

Ritz, P., and Schroeder, H. P. 1996. Life cycle cost analysis of a Storburn propane combustion toilet. *Proceedings of the International Conference on Cold Regions Engineering* 816–827.

Rivas, F. et al. 2006. Life cycle cost based economic assessment of active building envelope (ABE) systems. *Proceedings of AIAA/ASME/ASCE/AHS/ASC Conference on Structures, Structural Dynamics and Materials* 5534–5548.

Robbins, R. R. 1993. Life cycle costing factors for large valve regulated vs. flooded battery systems. *Proceedings of the Seventh International Power Quality Conference* 496–500.

Robinson, J. 1996. Plant and equipment acquisition: A life cycle costing case study. *Facilities* 14 (5–6): 21–25.

Robinson, T., and Smith, H. 1999. Cost and budget estimation for DoD ATE test program set acquisition and life cycle costs. *Proceedings of the IEEE Auto Test Conference* 469–477.

Rodriguez, G. A. R., and O'Neill-Carrillo, E. 2005. Economic assessment of distributed generation using life cycle costs and environmental externalities. *Proceedings of the 37th North American Power Symposium* 412–420.

Rooney, C., and Jackson, E. 1996. I.G. unit failure: A life cycle cost analysis. *Glass Digest* 75 (2): 44–52.

Roorda, O., McNeill, J. D., and Wright, M. 1996. Reducing the life cycle cost of swing check valves. *Proceedings of the International Pipeline Conference* 983–992.

Rose, M., and Sacks, I. 1995. Life-cycle cost implications of a system using bare SNF transfer. *Proceedings of the Annual International Conference on High Level Radioactive Waste Management* 340–342.

Rosenquist, G. et al. 2002. Consumer life-cycle cost impacts of energy-efficiency standards for residential-type central air conditioners and heat pumps. *ASHRAE Transactions* 108:619–630.

Rossegger, C. 2004. Variable medium speed pumps combine superior performance with reduced life cycle cost (LCC). *Proceedings of the Second International Symposium on Centrifugal Pumps: The State of the Art and New Developments* 89–102.

Roth, I. F., and Ambs, L. L. 2004. Incorporating externalities into a full cost approach to electric power generation life-cycle costing. *Energy* 29 (12–15): 2125–2144.

Rozis, N., and Rahman, A. 2002. A simple method for life cycle cost assessment of water sensitive urban design. *Proceedings of the 9th International Conference on Global Solutions for Urban Drainage* 1–11.

Rugg, D., and Fray, D. 2004. The role of net shape manufacture in reducing life cycle costs of gas turbine components. *Proceedings of the Cost-Affordable Titanium Symposium* 35–42.

Russo, J., and Ferro, S. 1994. Reducing life cycle costs. *Proceedings of the 8th Annual Conference of the Software Management Association* 5–6.

Rwelamila, P. D. 1996. Reducing life cycle costs of concrete structures: From routine testing to total quality management. *Proceedings of the 7th International Conference on Durability of Building Materials and Components* 781–782.

Saha, N., and Wang, M. 2000. A decision making framework for foundry sand using life cycle assessment and costing techniques. *Proceedings of the Annual Conference of the Canadian Society for Civil Engineering* 185–188.

Salem, O. et al. 2003. Risk-based life-cycle costing of infrastructure rehabilitation and construction alternatives. *Journal of Infrastructure Systems* 9 (1): 6–15.

Salem, O. M., and Ariaratnam, S. T. 1999. Infrastructure management: Decision support and life cycle cost analysis. *Proceedings of the CSCE Annual Conference* 349–358.

Sandberg, A., and Stromberg, U. 1999. Grippen: With focus on availability performance and life support cost over the product life cycle. *Journal of Quality in Maintenance Engineering* 5 (4): 325–334.

Sandberg, M., Boart, P., and Larsson, T. 2005. Functional product life-cycle simulation model for cost estimation in conceptual design of jet engine components. *Concurrent Engineering: Research and Applications* 13 (4): 331–342.

Sanitz, R., and Bitter, P. 1990. Control of life cycle cost through integrated logistic support. *Proceedings of the 7th International Conference on Reliability and Maintainability* 664–668.

Sarma, K. C., and Adeli, H. 2002. Life-cycle cost optimization of steel structures. *International Journal for Numerical Methods in Engineering* 55 (12): 1451–1462.

Sarma, V., and Karkhanis, S. 1997. System life cycle cost minimization—A simulation approach. *Proceedings of the Third International Conference on Modeling and Simulation* 132–137.

Sawase, K. 1991. Life-cycle cost study of co-generation systems using aquifer thermal energy storage. *Quarterly Report of the Railway Technical Research Institute* 32 (2): 116–122.

Schor, A. L., Leong, F. J., and Babcock, P. S. 1989. Impact of fault-tolerant avionics on life-cycle costs. *Proceedings of the IEEE National Aerospace and Electronics Conference* 1893–1899.

Schuh, G., Kubosch, A., and Leffin, T. 2004. Life-cycle costing in mold and die industry. *Proceedings of the 11th European Concurrent Engineering Conference* 10–13.

Schumaker, C. W., and Kankey, R. D. 1989. Life cycle cost management: The long term view. *Proceedings of the IEEE 1989 National Aerospace and Electronics Conference* 1221–1225.

Seiter, C. 1997. Advanced steam power plant concepts with optimized life-cycle costs: A new approach for maximum customer benefit. *Proceedings of the International Exhibition & Conference for the Power Generation Industries* 98.

Sekhar, S. C., Cher, T., and Kenneth L. 1998. On the study of energy performance and life cycle cost of smart window. *Energy and Buildings* 28 (3): 307–316.

Sellers, D. A. 2005. Rightsizing air handlers for lowest life-cycle cost. *HPAC Engineering* 77 (2): 26–34.

Seo, K. 2006. A methodology for estimating the product life cycle cost using a hybrid GA and ANN model. *Proceedings of the International Conference on Artificial Neural Networks* 386–395.

Seo, K. K., Park, J. H., Jang, D. S., and Wallace, D. 2002. Approximate estimation of the product life cycle cost using artificial neural networks in conceptual design. *International Journal of Advanced Manufacturing Technology* 19 (6): 461–471.

———. 2002. Prediction of the life cycle cost using statistical and artificial neural network methods in conceptual product design. *International Journal of Computer Integrated Manufacturing* 15 (6): 541–554.

Shaw, M. 2001. Medium-pressure UV reduces life cycle costs. *Water and Wastewater International* 16 (6): 27–28.

———. 2002. Life cycle costs reduced with medium pressure UV. *Water Services* 105 (1): 18–19.

Shen, Z., and Smith, S. 2004. Optimizing the functional design and life cycle cost of mechanical systems using genetic algorithms. *Transactions of the North American Manufacturing Research Institute of SME* 32:295–302.

———. 2006. Optimizing the functional design and life cycle cost of mechanical systems using genetic algorithms. *International Journal of Advanced Manufacturing Technology* 27 (11–12): 1051–1057.

Sherman, S., and Hide, H. 1992. Life cycle costing system for rolling stock. *Mechanical Engineering* 36 (1): 25–26.

———. 1992. Life cycle costing system for rolling stock. *Proceedings of the International Conference on Computer Aided Design, Manufacture and Operation in the Railway and Other Advanced Mass Transit* 25–33.

———. 1995. Life-cycle costing in support of strategic transit vehicle technology decision: Hamilton street railway looks to the future. *Transportation Research Record* 1496:59–67.

Shih, L. 2005. Evaluating eco-design projects with 3D-QFDE method and life cycle cost estimation. *Proceedings of the Fourth International Symposium on Environmentally Conscious Design and Inverse Manufacturing* 722–723.

Shimakage, T., Wu, K., Kato, T., Okamoto, T., and Suzuoki, Y. 2003. Life-cycle-cost comparison of different degradation diagnosis methods for cables. *Proceedings of the 7th International Conference on Properties and Applications of Dielectric Materials* 990–993.

Shonder, J. A., Martin, M. A., McLain, H. A., and Hughes, P. J. 2000. Comparative analysis of life-cycle costs of geothermal heat pumps and three conventional HVAC systems. *ASHRAE Transactions* 106:551–560.

Shore, B. 1996. Bias in the development and use of an expert system: Implications for life cycle costs. *Industrial Management and Data Systems* 96 (4): 18–26.

Shropshire, D., and Feizollahi, F. 1995. Life cycle cost estimation and systems analysis of waste management facilities. *Proceedings of the International Conference on Radioactive Waste Management and Environmental Remediation* 175–178.

Singh, D., and Tiong, R. L. K. 2005. Development of life cycle costing framework for highway bridges in Myanmar. *International Journal of Project Management* 23 (1): 37–44.

Sivill, T. E., Stoddard, D. N., Smith, T. H., and Roesener, W. S. 1993. Use of life-cycle cost estimates in the evaluation of proposed waste-treatment facilities. *Proceedings of the Technology and Programs for Radioactive Waste Management and Environmental Restoration Conference* 1797–1801.

Smith, R. L., and Kim, J. B. 1997. Life-cycle cost considerations for timber bridges. *Proceedings of the Structures Congress* 233–237.

Snyder, H. M. 1990. Life cycle cost model for dormant systems. *Proceedings of the IEEE National Aerospace and Electronics Conference* 1217–1219.

Soderstrom, M., and Nilsson, K. 1988. Simulation and life cycle cost optimization of industrial energy supply and energy use. *Proceedings of the Working Conference for Users of Simulation Hardware, Software and Intelliware* 77–82.

Soliveres, H., and Alquier, A. M. 1997. A particular aspect of DECIDE BID DECISION support system: Modeling of life-cycle processes and costs. *Proceedings of the IEEE International Conference on Systems, Man, and Cybernetics* 3609–3614.

Sone, S., and Takagi, R. 2004. The rapid transit system that achieves higher performance with lower life-cycle costs. *JSME International Journal, Series C* 47 (2): 539–543.

Songhurst, B. W., and Kingsley, M. 1993. Life-cycle cost reduction through designing for maintenance. *Proceedings of the Annual Offshore Technology Conference* 537–546.

Speck, R. P., and Herz, N. E. 2000. Impact of automatic calibration techniques on HMD life cycle costs and sustainable performance. *Proceedings of the SPIE Conference* 104–113.

Spector, R. B. 1989. Life cycle costs of industrial gas turbines. *Journal of Engineering for Gas Turbines and Power, Transactions of the ASME* 111 (4): 637–641.

Spence, G. 1989. Designing for total life cycle costs. *Printed Circuit Design* 6 (8): 14–15, 17.

Stahl, L., and Wallace, M. 1995. Life cycle cost comparison: Traditional cooling systems vs. natural convection based systems. *Proceedings of the International Telecommunications Energy Conference* 259–265.

Stalder, O. 2001. The life cycle costs (LCC) of entire rail networks: An international comparison. *Rail International* 32 (4): 26–31.

Stambler, I. 1997. New 190 MW design for 50 Hz aims for lowest possible life-cycle costs. *Gas Turbine World* 27 (5): 14–16, 18.

———. 1998. Utilities sponsor projects to improve reliability and reduce life cycle costs. *Gas Turbine World* 28 (5): 22–25.

Stanco, J., and Malesich, M. 1999. Reducing CV life cycle costs through process modeling and simulation. *Naval Engineers Journal* 111 (3): 359–370.

Stemetzki, G. A., Kuta, M. M., and Shepard, C. 2004. BOF hood life cycle cost improvement program. *Iron and Steel Technology* 1 (1): 54–67.

Stewart, M. G. 2001. Reliability-based assessment of ageing bridges using risk ranking and life cycle cost decision analyses. *Reliability Engineering and System Safety* 74 (3): 263–273.

Stone, K. W., Drubka, R. E., and Braun, H. 1994. Impact of Stirling engine operational requirements on dish Stirling system life cycle costs. *Proceedings of the Joint ASME/JSES/JSME International Solar Engineering Conference* 529–534.

Stoshi, I., Yasuto, I., and Tatsushi, H. 2005. High performance steel plates for shipbuilding: Life cycle cost reduction technology of JFE steel. *JFE Technical Report* (5): 16–23.

Stouffer, V., Hasan, S., and Kozarsky, D. 2004. Initial life-cycle cost/benefit assessments of distributed air/ground traffic management concept elements. *Proceedings of the AIAA 4th Aviation Technology, Integration, and Operations Forum* 783–797.

Strand, G. et al. 2000. Forecasting life-cycle costs. *Hart's E and P* 73 (8): 125–128.

Streicher, H., and Rackwitz, R. 2004. Time-variant reliability-oriented structural optimization and a renewal model for life-cycle costing. *Probabilistic Engineering Mechanics* 19 (1): 171–183.

Stump, E. J. 1988. An application of Markov chains to life-cycle cost analysis. *Engineering Costs and Production Economics* 14 (2): 151–156.

Su, C. T., and Chang, C. C. 2000. Minimization of the life cycle cost for a multistate system under periodic maintenance. *International Journal of Systems Science* 31 (2): 217–227.

Suleiman, T. et al. 1999. Practice, performance, and life cycle cost analysis of concrete pavement in Jordan. *Indian Concrete Journal* 73 (11): 687–692.

Sultan, N., and Groepper, P. H. 1999. Mobile satellite life cycle cost reduction: A new quantifiable system approach. *Proceedings of the Sixth International Mobile Satellite Conference* 246–251.

Suryawanshi, C. S. 2005. Life cycle cost model for concrete structures. *Proceedings of the IABSE Conference* 111–118.

Szeles, J. 1988. Life cycle cost (LCC) analysis of integrated circuits. *Proceedings of the 7th Symposium on Reliability in Electronics* 265–273.

Taehoon, H., Seungwoo, H., and Sangyoub, L. 2007. Simulation-based determination of optimal life-cycle cost for FRP bridge deck panels. *Automation in Construction* 16 (2): 140–152.

Takagishi, S. K. 1989. Electric power vs. petrol, methanol or gas: Life cycle cost comparison. *Electric Vehicle Developments* 8 (3): 77–78, 81.

Takahashi, T., Takeda, N., and Sogo, S. 2001. A study on minimization of life cycle cost for concrete structures using genetic algorithm. *Transactions of the Japan Concrete Institute* 23:157–164.

Takahashi, Y. et al. 2003. Life-cycle cost consideration in seismic risk management of a building. *Proceedings of the ASCE/SEI Structures Congress and Exposition* 1183–1190.

———. 2004. Life-cycle cost analysis based on a renewal model of earthquake occurrences. *Earthquake Engineering and Structural Dynamics* 33 (7): 859–880.

Tandon, M. K., and Seireg, A. A. 1992. Manufacturing tolerance design for optimum life-cycle cost. *Proceedings of the Manufacturing International Conference* 381–392.

Tao, Z. W., Ellis, J. H., and Corotis, R. B. 1992. Reliability-based life cycle costing in structural design. *Proceedings of the 6th International Conference on Structural Safety and Reliability* 685–686.

Tenca, P., and Lipo, T. A. 2005. Conversion topology for reducing failure rate and life-cycle costs of high-power wind turbines. *Proceedings of the 43rd AIAA Aerospace Sciences Meeting and Exhibit* 111–122.

Teo, E. et al. 2005. Maintenance of plastered and painted facades for Singapore public housing: A predictive life cycle cost-based approach. *Architectural Science Review* 48 (1): 47–54.

Thompson, P. D. 2004. Bridge life-cycle costing in integrated environment of design, rating, and management. *Transportation Research Record* 1866:51–58.

Tighe, S. 2001. Guidelines for probabilistic pavement life cycle cost analysis. *Transportation Research Record* 1769:28–38.

Titus-Glover, L., Hein, D., Rao, S., and Smith, K. L. 2006. Impact of increasing roadway construction standards on life-cycle costs of local residential streets. *Transportation Research Record* 1958:45–53.

Tozer, R., and James, R. 1997. Thermo economic life-cycle costs of absorption chillers. *Building Services Engineering Research & Technology* 18 (3): 149–155.

Treidler, B., Lucas, R., Modera, M. P., and Miller, J. D. 1996. Impact of residential duct insulation on HVAC energy use and life-cycle costs to consumers. *ASHRAE Transactions* 102 (1): 881–891.

Tuluca, A., and Heidell, J. 1990. Minimum life-cycle cost analysis of residential buildings for PC-based energy conservation standards. *ASTM Special Technical Publication* 1030:587–596.

Tupper, K., and Kreider, J. F. 2006. Life cycle impacts and external costs for various hydrogen pathways. *Proceedings of the ASME International Solar Energy Conference* 251–260.

Tutterow, V., Hovstadius, G., and McKane, A. 2001. Going with the flow: Life cycle costing for industrial pumping systems. *Proceedings of the ACEEE Summer Study on Energy Efficiency in Industry* 441–449.

Tzemos, S. 1990. Transportation cask life cycle cost uncertainty analysis. *Proceedings of the 1st Annual International Transportation Meeting on High Level Radioactive Waste Management* 1059–1065.

Uchida, K., and Kagaya, S. 2006. Development of life-cycle cost evaluation model for pavements considering drivers' route choices. *Transportation Research Record* 1985:115–124.

Ugwu, O. O., Kumaraswamy, M. M., Kung, F., and Ng, S. T. 2005. Object-oriented framework for durability assessment and life cycle costing of highway bridges. *Automation in Construction* 14 (5): 611–632.

Usher, J. S., and Whitfield, G. M. 1993. Evaluation of used-system life cycle costs using fuzzy set theory. *IEE Transactions* 25 (6): 84–88.

Vacek, R. M., Hopkins, M., and MacPherson, W. H. 1996. The development, demonstration and integration of advanced technologies to improve the life cycle costs of space systems. *Proceedings of the IEEE Aerospace Applications Conference* 217–225.

Val, D. V. 2005. Effect of different limit states on life-cycle cost of RC structures in corrosive environment. *Journal of Infrastructure Systems* 11 (4): 231–240.

Val, D. V., and Stewart, M. G. 2003. Life-cycle cost analysis of reinforced concrete structures in marine environments. *Structural Safety* 25 (4): 343–362.

Van Mier, G. P. M., Sterke, C. J. L. M., and Stevels, A. L. N. 1996. Life-cycle costs calculations and green design options: Computer monitors as example. *Proceedings of the IEEE International Symposium on Electronics and the Environment* 191–196.

Van Noortwijk, J. M. 2003. Explicit formulas for the variance of discounted life-cycle cost. *Reliability Engineering and System Safety* 80 (2): 185–195.

Verduzco, L. E., and Duffey, M. R. 2006. Modeling the financial and social life cycle costs of hydrogen-based systems. *Fuel Cell* 6 (4): 38–40.

Verduzco, L. E., Duffey, M. R., and Deason, J. P. 2007. H2POWER: Development of a methodology to calculate life cycle cost of small and medium-scale hydrogen systems. *Energy Policy* 35 (3): 1808–1818.

Verho, P. et al. 2006. Applying reliability analysis in evaluation of life-cycle costs of alternative network solutions. *European Transactions on Electrical Power* 16 (5): 523–531.

Veshosky, D., and Nickerson, R. L. 1993. Life-cycle costs versus life-cycle performance. *Better Roads* 63 (5): 33–35.

Vickery, P. J., and Twisdale, L. A. 1996. Reducing the vulnerability of transmission lines in hurricane regions by choosing minimum life cycle cost designs. *Proceedings of the Conference on Natural Disaster Reduction* 245–246.

Vipulanandan, C., and Pasari, G. 2005. Life cycle cost model (LCC-CIGMAT) for wastewater systems. *Proceedings of the Pipeline Division Specialty Conference on Optimizing Pipeline Design, Operations, and Maintenance in Today's Economy* 740–751.

Vivona, M. A. 1994. Audit environmental processes using life cycle costs. *Hydrocarbon Processing* 73 (8): 4.

Voigt, K. A. 2003. What is the real cost? [Test program set development life cycle costing]. *Proceedings of the IEEE Systems Readiness Technology Conference* 679–786.

Von Matern, S. 1992. Life cycle costing: Evaluation of method and use for stainless steel applications. *Proceedings of the Conference on Applications of Stainless Steel* 537–545.

Vorarat, S., and Al-Hajj, A. 2004. Developing a model to suit life cycle costing analysis for assets in the oil and gas industry. *Proceedings of the SPE Asia Pacific Conference on Integrated Modeling for Asset Management* 247–251.

Waghmode, L. Y., Birajdar, R. S., and Joshi, S. G. 2006. A life cycle cost analysis approach for selection of a typical heavy usage multistage centrifugal pump. *Proceedings of 8th Biennial ASME Conference on Engineering Systems Design and Analysis* 865–873.

Wang, E. 2005. Infrastructure rehabilitation management applying life-cycle cost analysis. *Proceedings of the ASCE International Conference on Computing in Civil Engineering* 1821–1830.

Wang, K. H., and Sivazlian, B. D. 1997. Life cycle cost analysis for availability system with parallel components. *Computers & Industrial Engineering* 33 (1–2): 129–132.

Wang, R. 1992. Research on aircraft life cycle cost reduction. *Journal of Aerospace Power/ Hangkong Dongli Xuebao* 7 (3): 291–292.

Warren, J. L., and Weitz, K. A. 1994. Development of an integrated life-cycle cost assessment model. *Proceedings of the IEEE International Symposium on Electronics & the Environment* 155–163.

Weber, W., and Fischer, M. 2005. Integrated logistics support: Product life cycle management: Controlling availability and costs. *News from Rohde and Schwarz* 45 (187): 56–57.

Weller, G. C., and Caunce, B. R. J. 1993. New distance relays reduce life cycle costs. *GEC ALSTHOM Technical Review* 12:55–62.

Wen, Y. K., and Kang, Y. J. 1997. Design based on minimum expected life-cycle cost. *Proceedings of the U.S.–Japan Joint Seminar on Structural Optimization* 192–203.

———. 1997. Optimal seismic design based on life-cycle cost. *Proceedings of the International Workshop on Optimal Performance of Civil Infrastructure Systems* 194–210.

Went, B. 2005. Life cycle costing—The integrated approach to management. *Proceedings of the 3rd International Conference on Water and Wastewater Pumping Stations* 89–105.

Westkaemper, E., and Osten-Sacken, D. V. D. 1998. Product life cycle costing applied to manufacturing systems. *Annals of the CIRP—Manufacturing Technology* 47 (1): 353–356.

Weyers, R., and Goodwin, F. E. 1999. Life-cycle cost analysis for zinc and other protective coatings for steel structures. *Transportation Research Record* 1680:63–73.

Whiteley, L., and Tighe, S. 2005. Incorporating variability into life cycle cost analysis and pay factors for performance-based specifications. *Proceedings of the 33rd CSCE Annual Conference* TR-121-1–TR-121-10.

Whiteley, L., Tighe, S., and Zhang, Z. 1940. Incorporating variability into pavement performance, life-cycle cost analysis, and performance-based specification pay factors. *Transportation Research Record* 1940:13–20.

Wies, R. W., Johnson, R. A., and Agrawal, A. N. 2005. Life cycle cost analysis and environmental impacts of integrating wind-turbine generators (WTGs) into stand-alone hybrid power systems. *WSEAS Transactions on Systems* 4 (9): 1383–1393.

Wilkinson, V. K. 1990. Life cycle cost analyses of government production and services. *Proceedings of the International Industrial Engineering Conference* 111–116.

Winkel, J. D. 1996. Use of life cycle costing in new and mature applications. *Proceedings of the NPF/SPE European Production Operations Conference* 239–248.

Wong, N. H. et al. 2003. Life cycle cost analysis of rooftop gardens in Singapore. *Building and Environment* 38 (3): 499–509.

Wonsiewicz, T. J. 1990. Life cycle cost analysis discount rates and inflation. *Proceedings of the ASCE International Conference on Pipeline Design and Installation* 639–648.

Woodward, D. G. 1997. Life cycle costing: Theory, information acquisition and application. *International Journal of Project Management* 15 (6): 335–344.

Woud, J. K., Smit, K., and Vucinic, B. 1997. Maintenance program design for minimal life cycle costs and acceptable safety risks. *International Shipbuilding Progress* 44 (437): 77–100.

Wu, K. et al. 2004. Life-cycle-cost assessment and correlation between degradation-diagnosis parameter and degradation degree. *Proceedings of the IEEE International Conference on Solid Dielectrics* 611–614.

Yan, X., and Gu, P. 1995. Assembly/disassembly sequence planning for life-cycle cost estimation. *Proceedings of the ASME International Manufacturing Engineering Division Conference* 2–2:935–956.

Yatomi, M. et al. 2004. An approach for cost-effective assessment in risk-based maintenance as a life-cycle maintenance (LCM) model. *Proceedings of the ASME Conference on Risk and Reliability and Evaluation of Components and Machinery* 41–46.

Yi, S. et al. 2003. Practical life-cycle-cost effective optimum design of steel bridges, life-cycle performance of deteriorating structures. *Proceedings of the Conference on Life-Cycle Performance of Deteriorating Structures: Assessment, Design and Management* 328–334.

Zackrison, H. B. 1991. How to reduce life cycle operating costs with A&E value engineering. *SAVE Proceedings* (Society of American Value Engineers) 26:143–153.

Zaganiaris, A. et al. 1992. Life cycle costs and economical budget of optical and hybrid access networks. *Proceedings of the 5th Conference on Optical/Hybrid Access Networks* 7.01.01–7.01.08.

———. 1993. A methodology for achieving life cycle costs of optical access networks from RACE 2087/TITAN1. *Proceedings of the Eleventh Annual Conference on European Fiber Optic Communications and Networks* 136–141.

Zaghloul, S. et al. 2004. Effect of positive drainage on flexible pavement life-cycle cost. *Transportation Research Record* 1868:135–141.

Zaghloul, S. M. 1996. Effect of poor workmanship and lack of smoothness testing on pavement life-cycle costs. *Transportation Research Record* 1539:102–109.

Zapata, J. M. 1994. Reducing life-cycle costs in ATE technology insertion. *Proceedings of the IEEE Systems Readiness Technology Conference* 439–442.

Zarembski, A. M. 1988. M/W first costs vs. life cycle costs. *Railway Track and Structures* 84 (9): 9–10.

Zayed, T. M. et al. 2002. Life-cycle cost analysis using deterministic and stochastic methods: Conflicting results. *Journal of Performance of Constructed Facilities* 16 (2): 63–74.

———. 2002. Life-cycle cost based maintenance plan for steel bridge protection systems. *Journal of Performance of Constructed Facilities* 16 (2): 55–62.

Zhang, T. I., and Kendall, E. 1999. Agent-based information gathering system for life cycle costs. *Proceedings of the 1st Asia–Pacific Conference on Intelligent Agent Technology Systems, Methodologies, and Tools* 483–487.

Zhang, Y., and Gershenson, J. K. 2003. An initial study of direct relationships between life-cycle modularity and life-cycle cost. *Concurrent Engineering Research and Applications* 11 (2): 121–128.

Zhang, Y. C. et al. 2005. Life-cycle cost analysis of bridges and tunnels. *Proceedings of the Construction Research Congress* 257–265.

Zhe, S., and Smith, S. 2006. Optimizing the functional design and life cycle cost of mechanical systems using genetic algorithms. *International Journal of Advanced Manufacturing Technology* 27 (11–12): 1051–1057.

Zhi, H. 2000. Simulation analysis in project life cycle cost. *Cost Engineering* 35 (12): 13–17.

Zimmerman, K. A., Smith, K. D., and Grogg, M. G. 2000. Applying economic concepts from life-cycle cost analysis to pavement management analysis. *Transportation Research Record* 1699:58–65.

Zoeteman, A. 2003. Life cycle costing applied to railway design and maintenance: Creating a dashboard for infrastructure performance planning. *Advances in Transport* 14:647–656.

———. 2004. Optimizing the performance of railway systems: Life cycle costing for rail infrastructure managers. *Proceedings of the IEEE International Conference on Systems, Man, and Cybernetics* 4159–4164.

Index

A

Aircraft airframe maintenance cost drivers, 109–110
Aircraft cost drivers, 108–110
Aircraft life cycle cost, 105–107
Aircraft turbine engine life cycle cost, 108
American National Standards Institute, 6
American Public Power Association, 7
American Society for Quality Control, conference proceedings, 4
American Society of Civil Engineers, 6–7
American Society of Heating, Refrigeration and Air Conditioning Engineers, 7
American Society of Mechanical Engineers, 7
Analysis cost
   as software life cycle cost, 97
   software life cycle cost element, 98
Analytic models, software cost estimation, 99
Annual American Society for Quality Control Conference, proceedings, 4
Annual Canadian Society for Civil Engineering Conference, proceedings, 4
Annual Conference of the Urban and Regional Information Systems Association, proceedings, 4
Annual Offshore Technology Conference, proceedings, 4
Annual Petroleum and Chemical Industry Conference, proceedings, 4
Annual Reliability and Maintainability Symposium, proceedings, 4
Annual Reliability Engineering Conference for the Electric Power Industry, proceedings, 4

ANSI. *See* American National Standards Institute
Appliance life cycle costing, 122–123
Application areas, 28–29
Areas for evaluation, 31–32
ASCE. *See* American Society of Civil Engineers
ASHRAE. *See* American Society of Heating, Refrigeration and Air Conditioning Engineers
ASME. *See* American Society of Mechanical Engineers
ASQC. *See* American Society for Quality Control
Asset condition, experience, 78
Automobile life cycle cost, 113–114

B

Bathtub hazard rate curve, 146–147
Better Roads, 3
Blanchard, B.S., 5
Boussabaine, A., 5
"Bridge Life Cycle Cost Analysis," 5
Bridge life cycle costs, 119–120
Brown, R.J., 5
Building energy cost estimation, 120–122
   formula I, 120–121
   formula II, 121
   formula III, 121–122
   formula IV, 122
   formula V, 122
Building life cycle cost, 117–118
Bull, J.W., 5
Bus life cycle cost estimation model, 114

C

Canadian Society for Civil Engineering, conference proceedings, 4
Car life cycle cost, 113–114

197